Praise for *Honey and Venom*

"[Andrew Coté's] journey to urban beekeeping is brought to life as he recounts often funny or bizarre situations. . . . [An] informative and entertaining memoir."

—*The New York Times Book Review*

"*Honey and Venom* is *A Year in Provence* for our modern times. With the fate, and the hope, of our world resting in the hands of species experts, Coté's charming and poignant essay collection delivers on the entertainment and smarts required to make real change in how we look at our planet, and ourselves. From biology to romance and mystery, the hidden culture of urban beekeeping pushes the stories of these magical creatures front and center, revealing their importance to our ecology and to our own human condition."

—Andrew Zimmern

"A fascinating story of Coté's real-life experiences with bees, full of unexpected plot swerves that take him all over New York City and all around the world on bee-related adventures . . . I enjoyed this book enormously."

—Ian Frazier, author of *Travels in Siberia*

"A fourth-generation beekeeper, Andrew Coté writes with wit, insight, and great empathy for the imperiled honey bee. The seasonal rhythms of Coté's bees mirror and often juxtapose the frenzied chaos of the human world outside their hives. Full of strivers and slackers, givers and takers, *Honey and Venom* is a fascinating story of urban beekeeping, social climbing, and the enduring power of family."

—MOLLIE KATZEN, author of *Moosewood Cookbook*

"In *Honey and Venom*, the stories are more than just amusing misadventures of an urban beekeeper—they are a richly layered intersection of beekeeping, art, history, and culture."

—HILARY KEARNEY, author of
QueenSpotting and *The Little Book of Bees*

"A fun and informative read, *Honey and Venom* showcases Andrew Coté's many talents, both as a masterful beekeeper and a fine storyteller. His passion and dedication to the honey bee is as unwavering as it is inspiring."

—SUSAN SPUNGEN, author of
Open Kitchen: Inspired Food for Casual Gatherings

"Andrew Coté has left a valuable and indelible mark on urban beekeeping thanks to his tireless efforts in New York City. And even more impressively, he's made a global contribution by teaching beekeeping as a sustainable source of income to communities around the world. This book takes an enjoyable journey chronicling Coté's fascinating activities. Delivered with charm, humor, and his compelling insights about honey bees, people, and culture, *Honey and Venom* helps us understand how the little bee has made such a big impact on his life."

—HOWLAND BLACKISTON, author of *Beekeeping for Dummies*

"[*Honey and Venom*] will educate and entertain even the most bee-phobic reader. Thanks to this delightful memoir, readers will have a new appreciation for these complex insects and the humans who care for them."

—*Shelf Awareness*

"Coté shares his exploits as one of the best-known beekeepers in the community; he is routinely called in to consult on many interesting cases, and we, the lucky readers, get to hear some of them in this charming read. . . . Fascinating, not only for the beekeeping information, but also for the urban-wildlife interactions involving bees . . . a good companion for anyone contemplating apiculture."

—*Library Journal*

"Entertaining . . . In his often amusing, anecdotal memoir, *Honey and Venom*, Coté offers the latest buzz on keeping an apiary in the Big Apple. Everything is told with Coté's light touch and excellent comic timing. . . . This book is fun, a near perfect bee-ch book for the summer."

—Minneapolis *Star Tribune*

"[A] delightful debut . . . Honey farmers and urban naturalists will be buzzing about this one."

—*Publishers Weekly*

"A fourth-generation beekeeper, [Coté] has amassed a treasure trove of wisdom. While the text is sprinkled with some advice for novice beekeepers, *Honey and Venom* is best read as a memoir—and a sweet one at that."

—*Booklist*

Honey
and
VENOM

Honey and VENOM

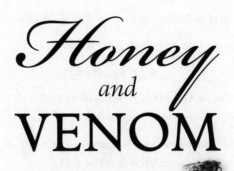

*Confessions
of an Urban
Beekeeper*

ANDREW COTÉ

BALLANTINE BOOKS | NEW YORK

2021 Ballantine Books Trade Paperback Edition

Copyright © 2020 by Andrew Coté

Published in the United States by Ballantine Books, an imprint of
Random House, a division of Penguin Random House LLC, New York.

BALLANTINE and the HOUSE colophon are registered trademarks of
Penguin Random House LLC.

Originally published in hardcover in the United States by
Ballantine Books, an imprint of Random House, a division of
Penguin Random House LLC, in 2020.

ISBN 978-1-5247-9906-9
Ebook ISBN 978-1-5247-9905-2

Printed in the United States of America on acid-free paper

randomhousebooks.com

2 4 6 8 9 7 5 3 1

Book design by Susan Turner

Drones don't have fathers, but I have a wonderful father, Norm,
who introduced me to the world of honey bees.

My precious queen bee, Yuliana, carried and delivered my perfect
little pollination project, Nobu, our impeccable brood.

I dedicate this book to my troika of sweetness and light.

Handle a book as a bee does a flower,
extract its sweetness but do not damage it.

—JOHN MUIR (1838–1914)

PROLOGUE

I bleed honey. It runs deep in my veins.

I am a fourth-generation beekeeper, intimately familiar with the world of honey bees and the aberrations of those who maintain them.

Early on, I honed my craft as a beekeeper on the array of hives that surrounded my boyhood home in Fairfield County, Connecticut. But today, aside from my apiaries in the suburbs and countryside surrounding Gotham, my turf includes skyscrapers, community gardens, ancient cemeteries, international territory, and other hidden pockets of New York City. My apiaries top, or have topped, some of New York's most iconic buildings and locales, from the Museum of Modern Art to the Waldorf Astoria. My bees buzz above the lawn of the United Nations and the cemeteries of Green-Wood in Brooklyn, Woodlawn in the Bronx, and St. Patrick's Old Cathedral down on Little Italy's Mulberry Street. They preside atop Industry City in Sunset Park and flitter over a ballet school near the Flatiron. They hover and dart above churches and synagogues, secondary schools and restaurants. The apiary I installed and tend above the seventy-second floor of a hotel near Central Park is the highest in the world, though admittedly not the most productive. While it is certainly a thrill for me to work with these industrious, extra-high-flying honey bees* at such lofty altitudes, I love working with those at ground level just as much—although I do envy the spectacular views of the city dwellers. All of these bees serve restaurants, grocery stores, my own honey stand, and, of course, first and foremost, themselves.

Beekeeping—and for beeks like me, urban beekeeping—is a passion for those who practice it. Worldwide, more and more people are drawn to this bewitching and satisfyingly messy work for environmental and conservation reasons, for the joy of producing something on their own, and because honey is, well, delicious.

One thing I love about beekeeping is how it brings people together in combinations one could not otherwise imagine. Because of it, I've cultivated the most unexpected acquaintances and friendships both close to home and abroad. I've worked with Parisian beekeeper Marie Laure Legroux on hives that have been on the grounds of the Luxembourg Gardens since 1856 (though Marie has not been there quite so long), and

---

* In his 1956 *Anatomy of the Honey Bee,* Robert Snodgrass writes in part: "We have in entomology a rule for insect common names that can be followed. It says: If the insect is what the name implies, write the two words separately; otherwise run them together." Thus we have such names as house fly, blow fly, and robber fly contrasted with dragonfly, caddicefly, and butterfly, because the latter are not flies, just as an aphislion is not a lion and a silverfish is not a fish. The honey bee is an insect and is preeminently a bee; "honeybee" is equivalent to "Johnsmith." Plus, who wants to argue with a guy named Snodgrass?

others with Nicolas Geant atop the Grand Palais des Champs-Élysées and Notre-Dame (the three atop Notre-Dame, it should be noted, survived the tragic fire that gutted the building in 2019). I've harvested from log hives in rural Samburuland in Kenya, helped prepare bees for overwintering in small Moldovan villages, relocated apiaries in immediate post-earthquake Port-au-Prince, Haiti, and helped Fijian beekeepers increase their profitability through targeted marketing of honey to honeymooners. Friendships were often forged along with those endeavors, many of which continue to this day.

Thoreau says that keeping bees is like directing sunbeams. Cupid supposedly dipped his arrows in honey before he set them flying. My favorite may be the words of Victor Hugo, who wrote "Life is the flower for which love is the honey." I think the Frenchman was onto something there. I adore these magical little creatures—from my rusticating hives in the countryside to their cosmopolitan cousins atop Manhattan high-rises. I love my bees and I love how I came to be a beekeeper.

I learned beekeeping from my father, Norm, who is distinctly humble and soft-spoken. A wiry fellow who as of this writing is in the latter portion of his eighth decade, he was a navy man poised on a ship off the coast of Cuba during the Cuban missile crisis. After his days on the water dried up, he turned to fire, serving for more than three decades as a lieutenant in the Norwalk fire department. In the days following the attacks of 9/11, he was part of a dogged rescue and recovery team digging fruitlessly in the pile of rubble that was once the World Trade Center. But like many men of action, Norm's manner is calm, patient, and distinctive.

His mother, Aldea, grew up on a farm with honey bees in Quebec, where she, along with her siblings and father, managed colonies of the nectar-gathering empresses of the sky. When she moved to the United States, she left the bees and the farm behind, giving rise to a brief period of dormancy for beekeeping in our lineage until Norm picked up the baton—or the bee smoker and hive tool. Whatever else we are, both my brother, Mike, and I are also now beekeepers, as are my brother's children; I hope mine will be as well. Through our charity, a 501(c)(3), Bees Without Borders, my father and I have traveled the world as a team carrying on the family tradition, teaching, listening, and working with bees while navigating our own modest bee business at home. And beekeeping for the beekeeper is a lot like it is for the honey bees themselves, or at least can be. Overlapping generations working together. Cooperative brood care. Division of labor. Mutual love of honey. We all have a lot in common.

In New York City, a place where passion and competition abound, there were only a couple of dozen beehives registered when beekeeping became legal in 2010 (although many more went unregistered). About a decade later, there are hundreds of registered beehives in the city with an estimated hundreds more unregistered. While most, myself included, would say that more beehives is a good thing, the zeal for beekeeping has led to a situation that any New Yorker, bee enthusiast or not, will find unsurprising: The business of beekeeping can be cutthroat. Many would correctly describe the act of beekeeping itself as humbling, serene, instructive, and even meditative. But it can also be rife with rivalries and animosity, turf wars and hostile takeovers. Sheepishly, I confess to once—but happily no longer—being among the most guilty of those megalomaniacs; there's only so much real estate for the bees, and there's a finite supply of nectar where I practice the ancient art. At long last the bees have taught me humility. Still, from time to time there are conflicts. The honey of a crowded hive is defended by a thousand stings. Or, at least, so say the *Olney Hymns*.

For all the competition, occasional bouts of unpleasantness, and stings—from bees and humans alike—there are many things I very much enjoy about being part of an urban beekeeping community. Living in Manhattan, I've had occasional brushes with celebrities, which are nearly always exciting and good fodder for stories—like the time I met and had a friendly chat with Denzel Washington, or when Natalie Portman walked off smiling, clutching a jar of my whipped honey. Aziz Ansari, the world-famous comedian, saw no shame in pragmatically bargaining five dollars off his purchase, and it was exciting when Uma Thurman, herself a beekeeper at the time, bought one of our honeycombs.

Setting up shop and hawking honey at the Union Square Greenmarket in lower Manhattan has introduced me to many intriguing people, including famous chefs, actors, and artists. Padma Lakshmi regularly comes by, and we talk about bees and life. Martha Stewart, whose bees my father tended for twenty years, comes by on occasion. Alec Baldwin and I have discussed the trials and tribulations of being a father later in life. David Schwimmer and I laughed about our uncanny resemblance when we were in our twenties, "which I hope was a compliment to us both," he said, certainly joking, since neither of us were pinups.

Hugh Jackman and I conversed for ten minutes without

my knowing who he was (nor, to be fair, he I); he was in the company of celebrity chef Jean-Georges Vongerichten and, the market being a foodie mecca, I thought that all of the hangers-on and cameras were there for Chef Jean-Georges. Hugh and I talked about distinct Australian honey varieties, like Leatherwood from Tasmania, and how on Hog Bay, Kangaroo Island, the Ligurian Bee Sanctuary continues to breed the last remaining pure Ligurian honey bee stock in the world. As he departed, having purchased a bottle of buckwheat honey, I asked him what he did for a living. "I do some acting, singing. . . . I'm doing something on Broadway now," he replied, very matter-of-factly and without a trace of amusement at my ignorance of his identity. "Excellent! Good luck with that!" I told him sincerely. When they were able to speak again, the two women who work with me informed me that I had been speaking with a megastar best known at the time for portraying Wolverine in the X-Men series.

On a more long-lasting level, thanks to beekeeping, I have made friends whom I cannot imagine otherwise ever getting to know. Most are people I had little else in common with—or so I thought—before the bees united us. Like Robert Deschak, a former marine captain with a degree in medieval literature, who once broke his collarbone while attempting to retrieve a swarm from a tree in front of the rectory of Saint Mary's Church on Grand Street; BJ Fredricks, a classical music singer, motorcycle rider, and chicken keeper, who's heavily involved in apitherapy and sustainability; and Valentina Ramirez, a Chilean-born woman who grew up in Tokyo and is currently finishing her doctorate, who astutely describes herself "like a T-Rex, with a big butt and short arms." Instagram influencer and native Manhattanite Eva Chen discovered my matcha-infused honey and told her followers about it, and now I have daily visitors thanks all to her. One of them, Michele Seelinger, a Jersey City–based Montessori teacher who loves honey bees and turtles, summed it up well: "[Beekeepers are] some of the kindest, friendliest NYC weirdos you'll ever meet!"

I have wanted to write this for a long time, but I had trouble figuring out how to make the time to sit and write a book on the divine flying creatures Khalil Gibran called "the messengers of love." Then it hit me like a bus. Rather, I was hit by a bus. I was delivering packages of honey bees to Brooklynites when, just entering Bed-Stuy at about midnight one rainy April evening, I was broadsided by an MTA bus. I was whisked off in an ambulance to a seedy south Brooklyn hospital in the middle of the night, covered in broken glass and my own blood. During my recuperation, the beginnings of this book took shape.

There is so much to tell! Most people know that, through their own alchemy, honey bees turn nectar into honey. But fewer know that they have five eyes, four wings, sort-of two stomachs, one for honey and one for digestion—though the honey stomach is not a digestive stomach, even though the name calls it a stomach. It is more like the crop of a chicken—and that their bodies are completely covered in hair, even their eyes and tongue. They have no ears but they feel vibrations with their feet and antennae. They can actually feel and respond to gravity. The drones (males) have no fathers but have grandfathers. Honey bees and dinosaurs coexisted, and the former have been around since the Cretaceous period. Honey bees have traveled aboard the space shuttle. They can be trained to detect bombs or cancer. They pollinate crops and help produce food for the planet's animals, including humans. They live atop the Paris opera house and on the roofs of buildings in Vilnius, Melbourne, and Kyoto. Honey bees have been part of the landscape in Iceland since the late 1980s. They have been used as weapons of war and a means to heal. I am all for the latter.

It has been a long haul, but despite the hard work involved, like working with bees, the effort of writing this book has made me happy, despite coming into contact with the occasional sting. In part for that reason, some names have been changed in this book. Ray Bradbury wrote that the bees' feet are dusted with the spices from a million flowers. If you've read this far I'll hope that you have chosen to enter the world of beekeeping through the portal of these pages. I hope that some of that happiness that the bees bring, like the precious powdered pollen being transferred by her magical and magnificent self from its origin of the flower to its new home in the hive, will settle on you.

Honey
and
VENOM

JANUARY

Such bees! Bilbo had never seen anything like them. "If
one were to sting me," he thought, "I should swell up as
big again as I am!"

—J.R.R. TOLKIEN, *The Hobbit*

Burnt goat meat on a stick. I ate it to help avoid getting
stung by bees. But, as I often do, I'm getting ahead of
myself.

Winter is the quiet season in beekeeping, which means
that January is a slow month for a beekeeper in the northeastern United States. The bees themselves do not hibernate;* but
they are dormant. They huddle around the queen to keep her

---

* As there are more than 20,000 types of bees in this great big world,
and something like 250 in New York City, please note that in these pages
when I refer to bees, unless otherwise noted, I refer to the honey bee—
Apis mellifera.

warm and to help her (and themselves) survive the frigid temperatures outside the hive in much the same manner as we gather indoors with family and friends around a crackling fireplace. Their cluster tightens and loosens in accordance with the outside temperature.

But even during the freezing months there may be the odd warm day, and that's when I make my rounds to check on as many of my beehives as I can manage. I may brush the snow from the front of the hives and hold a stethoscope up to the side of the wooden hive boxes to listen for the bees' hum within. I might employ a heat-seeking device to determine whether a colony is surviving and maintaining adequate temperatures in their cluster. Seeing a few dead bees in front of the hive is an encouraging sign—it means that live bees within, acting as undertakers, were recently active enough to toss out their sisters' little corpses. It's not as if I could do much of anything for my bees if they were faring poorly—if they've expired over the winter, they're gone. It is perfectly natural for the oldest bees to die in a colony on a continuous basis. During the winter these deaths are easier to see than during the rest of the season. Wondering if the bees are alive within is something akin to the question of Schrödinger's cat—the bees may be alive in the box, or they may be dead—but opening the box will certainly cause their demise, as all of their heat would escape. I cannot do much for them in the winter but hope. But I do worry about them.

Beyond fretting, there really are no everyday bee-related tasks to accomplish in our part of the world in January, so we beeks look forward to getting a little more sleep at this time of year. During spring and early summer I rise well before dawn to commence my rounds on the rooftops of buildings throughout the five boroughs of New York City and in the surrounding areas where I manage my apiaries. If not in the tri-state area at some beeyard or another, I am driving my pickup truck from place to place, stagnating in city traffic, sparring for parking spots, dealing with doormen, hauling equipment up and down fire escapes and makeshift ladders, handling reporters, and managing beekeeping apprentices. Since I'm not doing many of those things in January, I may sleep as blissfully and deliciously late as six A.M., catch up on non-beekeeping-related reading, or undertake a project that is normally impossible because of the demands imposed on me by my millions of four-winged mistresses.

Sometimes, if I need more to do in January, I'll follow the warm weather and go elsewhere to find beekeeping activity. Usually I do this through Bees Without Borders, the nonprofit 501(c)(3) organization that I founded with my father. BWB's mission is simple: to try to alleviate poverty via beekeeping endeavors. This organization allows me to combine four things that are personally important: philanthropy, travel, beekeeping, and education. Our operating budget has always been moderate; we are funded primarily through a percentage of the sales of our honey and value-added bee products such as pollen, royal jelly, propolis, and beeswax. Frugal New England beekeepers that we are, with our modest budget we've managed to make a small imprint on beekeeping communities in Nigeria, Cameroon, Iraq, Kenya, and Haiti to name a few. This way, aside from assisting people all over the world to gain a better income through working with honey bees, I don't have to wait for springtime to come to New York to play with bees and get my hands sticky and stung.

My family has been keeping bees since the 1800s. My paternal great-grandfather, Hector Laramée, kept bees in rural Quebec, near the Ontario border, on a small farm where they produced milk and honey. He learned these skills from his own father, who probably learned them from his. Hector worked alongside his wife and many children, including his daughter Aline and her younger sister Aldea, who was to become my grandmother.

One day Aline married a dapper, handsome, well-dressed man named Désiré Coté. Very much against the wishes of her kin, Aline moved to the United States with her new husband. The warnings of her clan unheeded, she soon found herself in an abusive relationship with a cruel and inexorable alcoholic. One day she wandered onto the railway tracks that passed beside the cemetery near the dilapidated house where they rented rooms, and was snuffed out by a Metro North train. Her two daughters watched the terrible incident. It was long rumored within the family that Aline had committed suicide to escape Désiré, but her two daughters firmly testified that it was an accident. In any event, she was gone.

At that cemetery, about ten feet from those tracks, where plots were the cheapest, Aline's widower bought a grave and had her interred with no marker. He took the settlement money from the railroad company—a small fortune back then, earmarked for his two young daughters—and spent it on himself, mostly on alcohol and clothes, since Désiré fancied

himself quite the dandy. Meanwhile his children were desti-
tute, hungry, and literally dressed in rags.

Immediately thereafter, between the two world wars,
Aldea moved to the United States to retrieve the four children
of her recently departed sister. Désiré preyed upon her as well,
putting her in the family way, and thus Aldea, a devout Catholic girl, married her former brother-in-law. In time she begot
her own brood of six children from her brother-in-law cum
husband.

The similarities of the only two households Aldea had
ever known north and south of the border mostly included
stark poverty and no indoor privy. Other than the usual hard-
ships of rearing a large brood under these conditions, and a
shared ethnic and cultural heritage, the households bore little
resemblance to each other. Whereas Aldea's father, Hector,
was a loving, hardworking farmer, Désiré squandered what
little money there was on his own wants rather than provide
food for his ten children. In turn, those children, including my
father, Norm, were forced to scavenge for food behind grocery
stores, telling the store employees they were looking for scraps
to feed the family rabbits. They would take these bits home
and separate the edible from the inedible and devour what
they could, sometimes giving the remnants to the rabbits, but
usually consuming it all themselves.

As a result of his circumstances, my father, growing up during a time of great economic prosperity in the United States
following the end of the Second World War, was severely under-
weight and regularly went to bed hungry, brushing his teeth with
his finger for a toothbrush and salt for toothpaste. Malnourished

herself, his mother, Aldea, lost her teeth and could not afford to replace them or get dentures for nearly two decades. She gummed her meat when she was lucky enough to get any.

As a young boy, my father spoke only French. In elementary school some classmates taught and encouraged him to call the teacher a "son of a bitch," which young Norm did, with no understanding of the meaning. For that he was kicked out of Winnepaugh Elementary School "until he can learn to speak English!" said the principal. Eventually he made his way back to the classroom, but not on a regular basis and not for long.

Aldea's move to New England began a thirty-year lapse in my immediate family's apiary activities until my father returned to his beekeeping roots in our backyard in Norwalk, Connecticut. At the time, my grandmother wasn't living close by, so Norm set out to teach himself what he needed to know during his downtime at his firefighting job—between car wrecks and house fires.

His small plot of land was no more than a quarter acre, and also held the small 1950s Cape Cod–style house in which we lived. Norwalk was a far cry from the dairy and honey farm that his ancestors had tended in Témiscaming, Quebec. Témiscaming, on the rare occasion it is thought of at all, is known as the administrative headquarters of the Algonquin Nation, and for winters that are harsh even by Canadian standards. Late in the season, before the snows, my great-grandfather would carry his many beehives down into the basement and place them on the dirt floor to wait out the brutal winters. Then he would fasten ropes between the house, the barn, and the outhouse so that nobody would get lost in the high snowdrifts when walking between the three.

I got involved with honey bees because they were a means to an end: I wanted to spend time with my father, a man who is at once tough and gentle, stern and loving. And always hardworking. Given that he had a full-time job, and beekeeping took up chunks of his time at home, helping him with the hives was a logical way to be in his company. As a firefighter, my father often smelled of smoke after work. He often smelled of smoke when he returned from his apiaries, too, the result of using a bee smoker, the oversized tin can–looking gizmo used to puff smoke into a beehive to distract and disorient the bees. This smoky smell, which I came to associate with my father, will always be pleasurable to me. The actual heavy, sweaty, and often difficult bee work never bothered me—in fact I still relish it—and any discomfort associated with it was unimportant relative to spending time with my dad. I couldn't have guessed it then, but as I shadowed my Quebecois father in his rickety pickup truck, my appreciation for the honey bee was cultivated and the family tradition was passed to yet another generation, from our father to my brother, Mike, and me. Nowadays Norm's grandchildren are in on the action, too.

I've always loved to travel, and bees have enhanced that passion. The majority of my twenties were spent living in Kyoto, where I earned my bachelor's degree in Japanese studies and studied martial arts six days a week. My father, it turns out, also loves to travel, though he wasn't able to do so until his children were grown. As a young man he did get to see some of the world courtesy of the U.S. Navy, but other than that, until his two boys were rousted from the family hive, he was too busy cobbling together a living for his growing boys, to whom he and my mother are still fiercely dedicated.

Once I hit the age of majority, my father and I started traveling the world together. On shoestring budgets, we've hiked England along the Coast to Coast Walk; biked from London to Marrakech long before GPS or the Internet would have made the ten-week trip easier; climbed Mount Kilimanjaro; and visited far-flung areas of Uganda. Way back in the late 1980s in Guatemala, we had the notion to visit local beekeepers rather than spend all our time hanging around backpackers' haunts and tourist spots. We figured meeting local farmers and sharing the day-to-day routine of their lives as much as they would allow—seeing their homes and fields, visiting their apiaries, and breaking bread with them—would give us a productive, and gratifying, way to go off-script.

We were generally welcomed by our fellow beekeepers, who seemed to relish hearing about our beekeeping practices as much as we enjoyed learning about theirs. We had a lot in common, as it turns out—most obviously a shared affection for the little four-winged creatures that transcended language barriers. We enjoyed the cultural exchange of comparing and sharing how we each lit a smoker; what we burned in it to confound the bees (burlap for us, dried elephant dung in Tanzania, willow bark in southern Russia, old shredded beekeeping equipment in Finland, mugwort in Korea and China, sumac bobs in Kalamazoo); how we harvested the honey and what applications it took on in our cuisines; and the manner in which we marketed the honey and products of the hive. As best we could, we discussed or demonstrated to one another the minutiae of our beekeeping lives, and we loved every minute of it. The local beekeepers seemed to find the exchange fulfilling, too. As for my father and me, we found the whole thing far more satisfying to our souls than visiting yet another colonial church.

That experience encouraged the two of us to continue taking beekeeping-centric trips. Soon we were seeking out as well as being sought out for projects by beekeeping clubs in places like Fiji and Zimbabwe, where club members asked for classes on topics such as queen rearing and basic bee biology. Sometimes they needed specific equipment, like a centrifuge to better harvest the honey, or a refractometer to measure the moisture content of the honey to see if it was suitable for harvesting without the fear of fermentation. These equipment requests raised the stakes somewhat. No longer just vacationers with an interest in bees, no longer just bystanders, we were becoming involved in local beekeeping communities who were looking to us for help.

And so we incorporated Bees Without Borders as a charitable organization. Which is how I found myself, in early January 2009, in Uganda, where I was about to learn one of the most important lessons of my beekeeping life. Namely, always pay attention to the rules of beekeeping as put forth by the bees. Even when you wish they weren't true. And the best way to eat overcooked, gristly goat meat is on a stick.

Norm, my friend Ben Gardner, and I were in a small village in the mountains along the Uganda-Congo border, helping about fifty or so local farmers hone their beekeeping skills.

We had traveled to East Africa at the request of a small agricultural collective that wanted help establishing a market for honey they did not have, from bees they had yet to appropriate, with skills that as yet eluded them. But they wrote an amazing proposal and hit all the right notes in what was to be a series of not-quite-completely-aboveboard applications coming out of a far-flung, poverty-stricken border community from people desperate to get any kind of help they could. We didn't exactly blame them for deceiving us—they wanted and needed help—we just had to figure out the best way to help them, despite their lack of transparency.

We traveled to Uganda via England. After a twelve-hour stopover in London that included warm beer and shepherd's pie, we made our way to Entebbe, and from there to Kampala. Our chock-full agenda was the result of months of preparation; we'd had to organize transportation, interpreters, guides, instructions for carpenters, models for the classes, and all sorts of materials and handouts for the students. We tried to think of everything. But particularly in Africa, even the best-laid plans need to be flexible, and we knew that things could change day by day, maybe even hour by hour. That old adage of hoping for the best but preparing for the worst comes in handy when you're traveling in a landlocked region where only about 20 percent of the population has access to electricity.

Corruption is an insidious problem in Uganda. In our experience, a project with a specific timeline is at the mercy of anyone with even an iota of authority. And power, when held, is wielded like a blunt instrument. It's often used to further the selfish and shortsighted interests of those in authority, and not the people or groups they were elected to represent. Yet despite the discomfort and corruption, it's always worth the effort and frustration to bring beekeeping skills to people in remote places. To see the immediate positive impact it has on the lives of the beekeepers, their families, and their community is humbling and changes all of our lives for the better.

Before leaving the Kampala area we visited an orphanage run by a Dane and well funded by Swiss bankers. The orphanage was clean, tidy, and well built—all thanks to the dedication and hard work of the tireless local staff. The paint was fresh, the floors were level, and the roofs didn't leak in the often-heavy rains. It was free from the pools of brackish water with which many compounds are plagued. The most heartening characteristic of the place, though, was its population—close to 150 healthy, well-fed children. They were cheerful and full of promise in a country that all too often offers neither cheer nor promise for the orphaned or anyone else. Norm, who has always loved and been loved by babies, was in heaven as the smaller kids gathered round and played with him. If he hadn't had his wife and grandchildren waiting for him at home he would have happily spent the six-month minimum time commitment volunteering there.

As we wandered from the orphanage's grounds we came to some enormous anthills, some six feet high and eight feet in diameter—basically the size of a Volkswagen Beetle. "Trample these anthills," the Dane told us, "and besides the ants getting into your pants, they will just rebuild them in a day." Returning to the property, we found the orphanage's dozen or so beehives, introduced by a volunteer who had since left the country. The staff hoped to increase the number of active hives—at this point there were only two—thereby increasing their honey yield, with the goal of having the children work the beehives and sell the honey as a means of income generation. Local beekeeper Lesster Leow, Singaporean by birth, was then managing the hives for the orphanage. He had already worked out an arrangement with some retailers in Europe to authenticate the purity of the honey and had established a niche market there that imported and sold the honey, with portions of the profit funneling back into the orphanage. It was a sweet deal.

"In 2005, before I was working with the orphanage, my first batch of EU-certified honey was selected to be supplied to a local supermarket in Kampala," he told me. "A Swiss banker spotted my little jar of honey sitting quietly on the shelf of that supermarket. He was searching for Ugandan products for Switzerland that could be sold to raise funds for the orphanage. We came to an agreement, and for the first time ever, the floodgates were opened for Ugandan honey to be exported to Switzerland." Lesster runs many other hives in Uganda as well as those at the orphanage. He sells some of his honey harvest from those hives locally, and he and his colleagues export the rest to Asia. He is nothing if not enterprising.

We visited the orphanage on New Year's Eve 2008. That evening, Norm watched fireworks from the balcony of our small hotel. Ben and I sat down below in the outdoor restaurant, drinking Nile beers and watching the same fireworks shot off the roofs of the upscale hotels. We drank, avoided the bevy of local prostitutes, and listened and watched as two young women belted out their renditions of "Stand by Me," punctuated by explosions overhead. There wasn't enough beer in the country to make that off-key caterwauling worth enduring, so Ben and I returned to our shared room for a few hours. At six A.M. Lesster picked us up for a long trek west in his SUV. No air-conditioning.

The journey was long, hot, and bumpy, but Lesster's company made it pleasant. In addition to his work with bees, he also had a small shop along a busy road where he sold locally made crafts such as carved wooden masks, handwoven raffia baskets, wall tapestries, and jewelry made from everything from bottle caps to tightly wound paper dipped in glue. Lesster sold only goods that were truly locally made, not the usual tourist trinkets passed off as local but in reality manufactured in Kenya, Indonesia, or China. By supporting local craftspeople, his shop directly benefited the communities surrounding Kampala.

During the marathon drive to our first destination, Kasese, we passed several coffin shops, rickety stores that displayed their simply made coffins on the side of the road. Life expectancy in Uganda is decades shorter than it is in the developed world, so there is, unfortunately, a large demand for coffins. In his late sixties at the time, Norm was already older than 98 percent of all Ugandans. Poverty, disease, and instability breed untimely deaths. Most Ugandans don't make it to their mid-fifties.

I often hear people who don't really know any better pontificating about how poor people from rural regions have a stronger connection to nature than people in wealthier, more industrialized countries do, and a greater respect for animals and the earth. My experience indicates otherwise. For instance, in Fairfield County, Connecticut, where I grew up, as well as in Manhattan, where I live now, there are psychologists specializing in assisting emotionally distraught dogs. While I am not against such a thing, since canines can surely suffer emotional trauma just as humans do, people who are chronically hungry do not generally have the luxury of worrying about whether a mule is beaten, let alone having emotional issues. They have bigger problems, like rampant disease and starvation. Beekeepers in deeply impoverished areas rarely have any passion or compassion for the bees, a sentiment that can perhaps be understood in context. These beekeepers often burn out the hives, kill all the bees, and take all the valuables—namely the honey. They do not smoke the bees out, allowing them to live; they burn them out.

Even the few who have had training with working hives (not feral colonies) will still, for the most part, pull all of the honey out of the hive rather than leave enough for the bees to survive. Then again, in a country where a sizable percentage of the population subsists on around a dollar a day, it can be difficult to convince a beekeeper not to grab all that he or she can, much like the country's leaders. Tomorrow isn't promised. Something or someone else might come along and steal the honey—a neighbor, an animal—and then there would be nothing. So some just grab what they can while they can, in these cases, to the detriment of the bees.

During the bumpy, nausea-inducing drive, we encountered baboons, kobs, massive warthogs, and huge garbage-eating storks. Certainly one thing can be said of Uganda: It is green—green in every direction as far as we could see. The lush natural beauty is spectacular. At some point we stopped for deep-fried tilapia from Lake George, and then we traveled the road that traversed Queen Elizabeth National Park for several hours. The lush natural beauty there is spectacular. In the middle of the park we met up with one of Lesster's apprentices, a local teenager named Benjamin who had a few dozen beehives sprinkled along a steep hillside and was starting to make his own living as a beekeeper, at least in part.

We crossed the equator on the way to Kasese. We crossed it more than once, in fact, as the road snaked north and south over the imaginary line. Though usually flying no more than three miles at a time, honey bees must travel the collective equivalent distance of two times around the equator, or fifty thousand miles, to gather enough nectar to produce a pound of honey. Norm, Ben, and I had a much easier time reaching our goal, stopping to take silly tourist photos of one another standing at the equator with one foot in either hemisphere. The road continued to snake west, on even more questionable roads, to what truly felt like past the middle of nowhere toward the far edge of nothingness. Once we reached Kasese, we bought some supplies, bade goodbye for the time being to Lesster, and headed even farther off the beaten path to the mountains near the border with Congo.

Two weeks later, Norm, Ben, and I had just wrapped up ten days of theoretical classroom work aimed at educating farmers on how to manage honey bees. We had been staying in a local hotel that provided one meal per day, and had heard one of our translators boast that "we have more than forty types of bananas in Uganda!" It was a fine brag and I had no idea if it was true, but we had indeed discovered that bananas were in some shape or form part of just about every Ugandan meal. Too bad for us, because when it comes to honey bees, bananas smell like death.

Honey bees have an advanced and complex pheromonal communication system rivaled by few other creatures. They possess varied scents secreted by the three castes of honey bee (queen, drone, and worker) that relay messages or encourage responses from the others in the colony. One of the two primary alarm pheromones, released via the Koschevnikov gland, located near the sting shaft, is made up of dozens of different compounds and is released when a bee stings another creature. This scent is emitted in order to communicate to nearby bees that there is danger in the area, raising an alarm to send any and all resources to join the fight to ward off or put down the threat. And this scent happens to smell like bananas.

So prior to visiting an apiary with about ten colonies of bees—which, in the case of this particular apiary in the mountains of western Uganda, meant as many as one million flying, stinging insects—to eat bananas is to smell like bananas is to invite honey bees to attack. For that reason, bananas should be avoided.

When a honey bee stings any mammal, she is essentially giving her life to defend her extended family, the colony, which may be why most people believe that honey bees can sting only once. But this is true only some of the time. Honey bees can sting other honey bees because of where they sting, in the membrane at the base of the wings, which lacks the tissue to pull the sting structure from the base of the stinging bee. If the recipient of her sting is a human or other mammal, more than likely the honey bee's stinger will imbed into her victim, with the barbs along her shaft gripping the flesh of the recipient. As the bee alights after plunging her stinger into the meat of her enemy, part of her abdomen and other bits will be torn off and left behind, and the bee will indeed die. If her victim is another insect, however, the honey bee may just live to sting another day.

So the punctured beekeeper may take some solace in knowing that, at the very least, the source of his or her pain is dead, or will soon be. (The honey bee, by the way, is the only type of bee to die as a result of stinging.) The pinprick of pain that a human experiences is barely felt in most instances, but the venom that's pumped into the flesh is a complicated mix of proteins that are a toxin.

So that banana-like alarm pheromone encourages other honey bees to sting again—and again and again—in the same area. That's why, on the morning that we were finally ready to take our students to the feral hives we planned to reclaim, I said no to banana mash for breakfast and yes to goat on a stick, as charred and as questionable as it looked. And that's why I belabored the point to my students: "Eating bananas on a day when visiting beehives is a bad idea," I told them repeatedly, and then some more.

Our students were from a few different villages in the general area. Many of them were widows whose husbands had died from HIV/AIDS, malaria, or one of the many conflicts in the region. For this series, Bees Without Borders had been invited by two nongovernmental organizations, or NGOs, set up for the betterment of the extended community. Our mission was to help farmers increase their crop yields through pollination and to teach them to cultivate honey as a cash crop that they could store and sell potentially even years in the future, if need be for whatever reason. Our by-now-beloved students found it particularly funny when I stole a line I use when hawking honey at the farmers' market—I told them that unlike marriage or children, honey never spoils.

In any case, we would be introducing beekeeping skills to some, and elevating those skills in others—although as it turned out, more the former than the latter. And it wasn't just aspiring beekeepers who attended. Since our classes included modest meals prepared by locals who were hired for that purpose, many people came for the free lunches. They were most welcome. Others were probably there to stare at the trio of *mzungu* (white people) who were daft enough to leave their comfortable homes in the United States and appear in this place. For still others it was likely just something different to do. No matter what their primary goal was, we enjoyed the lively company of our students. And lively they were.

From the very first day, they asked question after question about life in the United States. After a while we started to dedicate a short period of time at the end of each session to answering these non-bee-related questions.

"Have you met Obama?"

"Is everyone in America rich?"

"Are all Americans beautiful?"

From magazines and television, our students seemed convinced that there was no poverty in America and that everyone was good-looking. We assured them that we knew many people in America who were poor and unattractive, pointing to Ben as a perfect example of both.

The classes, though long, were popular. Some people walked for hours every day in order to attend the lectures. There were two interpreters present who worked hard to keep the information flowing, the questions and answers moving back and forth smoothly. The local NGOs were represented as well.

Sessions were held in a cinder-block building with holes in the walls for windows. A sheet of warped plywood painted black served as a blackboard. Desks and chairs, clearly meant for children, were occupied by adults. The hardworking women present were all barefoot, all garbed in colorful clothing made from the same few bolts of fabric. Several had babies swathed around their bodies, and at least two dozen small children, dressed in rags or otherwise undressed, played in the grass outside. The little ones peered into windows, climbed trees, or chased one another, all the while chattering and laughing ceaselessly. It was wonderful background music. Mothers nursed babies during the lectures.

The students' enthusiasm made our task joyful. Still, it was an uphill slog. As far as we could surmise, none of the students were literate, so we used photographs, drawings, and a lot of repetition to try to make our points as we progressed through the curriculum. We let the experience be interactive. Students drew diagrams. They acted out the various roles within the beehive. They pretended they were young (under three weeks) worker bees and described their tasks: taking care of the babies, tending to the queen, rebuilding honeycomb, and performing the duties of an undertaker, to name just a few. The assessments near the end of the theoretical material showed that most of them grasped the concepts quite well, and I was excited to move on to the practical side of the training—handling the bees themselves.

We had gone over specific safety instructions in the days leading up to our first ever apiary visit. The students, mostly women, knew to avoid drinking the night before, as we had always been taught that the scent of alcohol enraged bees; to have a good wash, and not to use any perfume, lotion, or strong-smelling shampoos. They were also told they would have to leave all babies and children behind in the classroom, where others would look after them.

And, perhaps most important of all, they knew to avoid any type of banana until after the visit. This last point was key as we didn't have enough veils for half of the students—there were many more than we had expected—and so people would need to take turns.

On the day of our planned visit to the hives I walked up a hill along a path worn in the grass, as I had every other morning since we'd been there, past a small church with a dozen or so handmade signs hanging in the courtyard that proclaimed VIRGINITY IS NOBLE and SAVE YOURSELF FOR MARRIAGE. I entered the classroom to find myself inhaling the same earthy body scents as I had for the previous two weeks; clearly, not much of the bathing we'd advised had happened. More dismaying was watching the students sitting in their usual seats gorging themselves on generous bowls of banana mash. And then strapping their babies to their bosoms and backs. Our repeated protests went unheeded. Still trying to find a good solution, we all set off for the walk to the apiary; it was so hot and humid that the sooner we left the better off we would be. After conferring during the hike, Norm, Ben, and I came up with a plan that we thought would allow us to proceed safely.

I had attempted to visit the apiary prior to this group trip, but I had been unable to persuade anyone to take me; what felt to me like a necessary site survey felt like an unnecessary trek to the would-be guides. So I wasn't sure exactly what to expect. I had been told that there were ten functioning beehives in two styles—top-bar hives and Langstroth hives[*]—that were located ten minutes down the road by foot. The actual beekeeper had left the area, or was unavailable, or had died—depending on which day and who I was talking to. The trip turned out to be a ninety-minute hike up and down a natural obstacle course. Having been in Uganda for a few weeks at this point, I wasn't surprised to find that the journey was ten times the projected time and distance. The road was more like a path. Sometimes wide enough for several of us to walk abreast, and sometimes narrow, overgrown with brush to the point that we would have to hunch over to get through the thicket single file.

Eventually, our guide directed us to walk off the trail toward the apiary. He didn't lead, he pointed. Perhaps he alone had listened to our instructions and warnings. For

---

[*] A top-bar hive is a style of beehive made from simplistic and easily found materials, developed in the early 1970s by a pair of Canadian PhDs from the University of Guelph, for use in East Africa. The boxes essentially look like small coffins and have fewer parts than the Langstroth hive. The Langstroth was developed in the 1850s by the Reverend Lorenzo Langstroth and is the most commonly used type of beehive in North America and perhaps the world.

whatever reason, bright fellow that he was, he wasn't going anywhere near the bees.

Our students, and the banana mash they consumed, gave us pause and we assessed the situation again. The honey bee has antennae that are used to smell, taste, and hear—but mostly to smell. Their antennae house an astounding number of odor receptors—drones having the highest number. Honey bees have no ears but use their antennae and their six legs to detect vibrations. Their wings stroke 11,400 times per minute, which means they fly faster than an average Ugandan runs. So, even though their thousands of lenses in their two compound eyes (which complement their other three eyes) render images blurry farther away than a couple of meters, not unlike pixels on a screen, the bees clearly interpret those pixel-like images in their brains. Thus they would have no trouble at all in locating our team and causing havoc.

Had I not been such a willful person myself, and had we not come so far in order to assess this apiary, I probably would have voted to postpone this event. I ought to have. But again we three instructors conferred and had a frank conversation with the interpreters. We told them the three of us would put on our veils, and, along with the two beekeepers we had recruited from the nearest village, would check out the situation and return for small groups of a dozen or so at a time, in order for there to be enough protective gear. We told them to make themselves comfortable in the shade, and impressed upon the interpreters in no uncertain terms that they were all to remain where they were until we returned for them. Then we started the trudge through the bushes and small growth and brush to the apiary one hundred or so yards away.

We arrived to find what was essentially an abandoned apiary with perhaps six live hives, two of which were hanging in the trees—a not uncommon occurrence in this vicinity, as it helps to keep the hives relatively safe from animals and insects. Several other hives were lying in half-decomposed states on the ground or half propped up on makeshift stands. It was far from ideal, but it was a start. We would work with what we had and move forward.

We had brought a small handheld video camera with us and began to do a short interview with one of the local beekeepers there in the apiary as we like to share these journeys with other beekeepers—those who have donated to our cause, and anyone interested, really. We were completely unaware that the entire group of aforementioned banana-reeking Ugandans had ignored our directive to stay put and instead had been stealthily making their way closer and closer to us. Stealthily, but not cautiously.

Before we even had a chance to fully gather our bearings, the behavior of the bees became much more aggressive. I had been looking through the viewfinder of the camera at the two local beekeepers when I heard a commotion and turned to see my father and Ben behind me energetically trying to get the dozens of curious new beekeepers-to-be to turn around and head back to safety. But it was too late. Their scent, the rumbling and vibration of their footsteps, or maybe their voices—or who knows what combination thereof—caught the attention of the bees and irritated them, and their response was immediate and definitive. They attacked.

The honey bee is *Apis mellifera*. *Apis* means "bee," and *mellifera* means "honey bearing," from the Latin *melli* for "honey" and *ferre*, "to bear." (Actually, honey bees are nectar bearing and not honey bearing, but we can forgive Carl Linnaeus for making the error back in 1756.) There are currently twenty-eight different subspecies of *Apis mellifera* recognized in the bee-keeping community. Most people have heard of Africanized bees, the so-called killer bees. Many people mistakenly consider the bees in Africa to be those Africanized bees, but that is not the case. In most of southern and eastern Africa, we deal with *Apis mellifera scutellata*. Though *scutellata* is certainly more aggressive than its cousins the *ligustica*—the type of bee that the majority of beekeepers in North America are likely to use—it is not a "killer bee." It was in the 1950s in Brazil that *Apis mellifera ligustica* (the Italian honey bee) and *Apis mellifera scutellata* (an East African lowland honey bee) were combined and the so-called Africanized or "killer bee" was born. The new subspecies went on to inhabit the majority of the South American continent and earn itself its vicious reputation. But this is all one interpretation. Many scientists believe that, thanks to their usurpation behavior, the Africanized honey bee is extremely close to the African (*scutellata*). In either case, they were not hospitable.

When I was at elementary school in the 1970s, it was common on the playground to hear that "the killer bees will arrive in five years!" or some such made-up time frame. Younger kids gasped as their elders, aged eight or nine, stoically shook their heads and secretly enjoyed the fear they were inspiring. While there are parts of the United States where these bees have indeed penetrated and changed the beekeeping industry somewhat, they have not yet settled into most parts of the country. This despite the seven B-list (bee-list?) films from that decade which, much like those playground fearmongers, spread the notion that killer bees would soon arrive en masse to kill us all.*

Still, in Uganda, those *Apis mellifera scutellata* were not gentle souls themselves and were certainly no slouches in the stinging department. Later Norm, Ben, and I compared notes, and we had the same impression: For whatever reasons, the group had ignored the safety instructions we thought we had impressed upon the translators (or perhaps the translators had not conveyed the instructions well, or at all) and had made its way through the brush to get a sneak preview. None, including women with babies strapped to them and a few elders, had donned the protective gear that sat in a haphazard pile back by the side of the so-called road. When the bees attacked, the crowd panicked—screaming, shouting, wildly waving their arms, and running. Unfortunately they ran in several different directions, which made it tough to lead them in an expeditious exit. Also, there is no such thing as graceful running in the sort of brush we were in. The ground was mined with holes and trip-wired with roots and vines. Since honey bees fly an average of fifteen miles

---

* The 1978 film *The Swarm*, which boasted a cast including Michael Caine, Richard Chamberlain, Henry Fonda (who was, in real life, a beekeeper, and produced honey he labeled as from Hank's Bel-Air Hive), Olivia de Havilland, and even Slim Pickens (who by that point had long outgrown his monicker), may be the best known and most ludicrous of all of them, though it has some pretty stiff competition in the ridiculous category from the likes of *Killer Bees* (1974), *The Savage Bees* (1976), and a gem from the pre-moon-landing days of 1966, *The Deadly Bees*. In one scene from *The Swarm*, Olivia de Havilland, a two-time Academy Award winner, was playing dead and covered in thousands of bees, when one live bee crawled up her nostril. Gravely she maintained her composure for the shot.

per hour, and a human under ideal circumstances can run an average of about half of that, the three of us knew we were in for a bad morning. And it might have been my imagination, but these bees seemed to actually fly ahead of us, then turn around and come at us face-to-face. They were menacing.

We three *mzungu* were wearing our bee jackets and veils, so we could have emerged mostly sting-free but for a few bees that would have found their mark on the tighter bits of clothing or up the pant legs. But almost immediately into the rumpus, my father tore off his jacket veil and placed it over the head and body of a child. I'm not sure whether we were more inspired by his selflessness or ashamed for not having thought of it ourselves, but Ben and I quickly followed suit with our own gear while grabbing more vulnerable people and pointing them in the direction of relative safety. One of our students was thinking clearly enough to go back for the additional veils that were mind-bogglingly left behind at the path, and we stuffed heads into them as we shuffled bodies toward safer ground.

The entire incident lasted only a few minutes before we managed to get everyone to relative safety at the original waiting point. Later we discovered that the camera I'd been using to video the interview with the local beekeeper was still running. Though the images were not distinct or steady—the camera was bouncing from a shoulder strap—the audio was compelling enough to give a clear idea of the sort of melee that had just taken place. On it my voice can be heard shouting "Cover your throat!" to my father, since I didn't know how many stings to the throat any of us could take out there, and there is a great deal of panicked shouting. Ben had been wearing a sort of handkerchief around his neck. While at the start of the day I teased him for looking like a fop in it, the joke was on me, as it served well to cover his mouth and nose once his veil was repositioned onto a young mother. The three of us were stung the most because we were in the lead, and therefore the first victims the bees found when darting toward our group, and we were also the last to leave, as we repeatedly returned to make sure everyone had made it out. It certainly could have been worse, though it did not seem so at the time.

We immediately saw to the babies and children, none of whom had taken more than a few stings, if any. Some of the adults had been hit multiple times, and they were checking one another for wayward bees in folds of clothing. A few of the students sweetly tried to be helpful by plucking dozens of stingers out of my face. What I knew but did not care about at that instant was that they were in fact injecting me with yet more venom, as the venom sack that is attached to the end of the hypodermic needle–like barbed stinger continues to pump poison into the flesh of the sting recipient. Grabbing and pulling out the stinger simply forces all of that venom into the skin at once. Some scientists have disputed this claim, stating that the majority of the venom enters the body immediately, and there is no difference whether the stinger is cut out with a knife edge or a fingernail, or is plucked out of the skin like the feathers of an emaciated chicken ready to be thrown into a pot over an open fire. Either way, a bee sting hurts. Dozens to the face hurt more.

Considering the setback, we decided that we would return the next day in smaller, more manageable groups, in full gear. Then, in a humbler, more somber mood, we plodded the hour and a half back to the classroom, where we resumed the lessons despite the defeat of the morning. In a short while lunch was prepared and served. I joined the others and feasted on a well-earned meal of banana mash, which I consumed with gusto, though my face pulsed with pain and I could feel my skin beginning to tighten around my eyes. At the same time, the joints of my fingers had swollen so much that my fingers looked like linked sausages and were becoming difficult to bend. Oddly, sitting in discomfort there, in the stifling heat and humidity, slowly eating a pile of room-temperature mush, I was quite happy. I found myself thinking of my own beehives atop their varied rooftops in Manhattan and Brooklyn, deep in their winter slumber. I hoped the bees within would still be alive despite the cold, hunkered together and vibrating and shimmying and moving in a specific rhythmic circle pattern in order to share the warmth of their bodies and transfer food to one another. I also reflected upon how lucky I was that I could spend time with these people, who were nearly completely at the mercy of nature for their survival, and yet had accomplished so much with so little. Knowing that it was honey bees and an affinity for them that somehow connected us all filled me with contentment in spite of the disastrous morning. I couldn't stop grinning; or rather, trying to grin, as my swollen face prevented it. I tried to communicate some of my feelings to Ben and my dad with my eyes. Then, with a mouth that barely moved, I asked, "Does it look bad?"

My father stopped spooning the gruel into his mouth and looked at me intently with his bright blue eyes. He leaned in close to me and said in a low, serious voice, "It doesn't look good." And then he laughed.

FEBRUARY

Everyone should have two or three hives of bees. Bees are easier to keep than a dog or a cat. They are more interesting than gerbils.

—SUE HUBBELL, *A Book of Bees*

As counterintuitive as it may seem, honey bees and penguins have something in common: Both huddle together for warmth. When Antarctic temperatures reach as low as 40 degrees below freezing, emperor penguins waddle their way to a central point and tuck themselves into a tight cluster to maintain a common core temperature. They shift and rotate from the outside of the bevy, where the brunt of the extreme cold is felt, to the center where the temperatures are far cozier. These tuxedo-wearing swimming birds continue to alternate with one another so as to share the burden of cold as long as the extreme temperatures require them to do so.

Back in the northeastern United States, honey bees, like their web-footed, fish-eating, better-dressed friends, form a tight cluster around her majesty the queen in order to survive the frigid temperatures outside the beehive. Her royal highness stays in the center and remains warm—her environs maintained at a consistent 95 degrees—while her daughters and few sons rotate, sharing food gathered from the outer perimeter of the cluster and warmth garnered from within. The oldest worker bees, born at the tail end of summer, die and are allowed to fall to the bottom of the hive.

I grew up thirty-five miles from New York City on the southern tip of penguin-free New England in the coastal blue-collar town of Norwalk, Connecticut, where my father's family settled in the 1930s. Once upon a time Norwalk was known for oysters, hat making, and as the last producer of rock candy in the United States. The oyster industry is slowly making a comeback after a century of steady decline. The rock candy industry, along with the hat-making business, has long since hit rock bottom.

Norwalk is surrounded by some of the wealthiest towns in the United States—Westport, New Canaan, Darien, and Wilton, as well as Greenwich just down the road—in what is referred to as the Gold Coast. Still, Norwalk continues to lag far behind its neighbors in terms of financial means. When I was growing up in the seventies, it was home to sanitation workers, police officers, firefighters, and other public works employees—basically, anyone whose job it was to serve the posh neighboring towns.

My own family is dug into Norwalk like ticks. My father is one of ten siblings, and my mother one of eight, so multiply that by spouses and children and we end up with quite a large litter. My father was a lieutenant for the Norwalk Fire Department for more than thirty years, and my mother worked for Southern New England Telephone, starting as a switchboard operator at sixteen. Our extended family has long been embedded in the city's police and fire departments, the schools, the library system, city hall, the now-defunct draft board, the department of transportation, and other municipal organizations. In other words, we've had a stake in the community for years. And despite Norwalk's population of about eighty thousand, we know most everyone, or we know someone who knows them.

During our elementary and middle school years, my brother, Michael, and I each had our own newspaper routes. This was back in the days when *The Hour,* then a locally owned and run newspaper, was delivered by children on bikes after school. Now the paper is owned by the Hearst empire and is delivered before dawn by adults in cars.

In our neighborhood, houses were modest and on small lots. We lived within walking distance of our elementary school and the sprawling, wooded, and hilly cemetery, where my paternal grandfather hid liquor bottles, kept rabbits, and dug graves, in that order of personal importance.

When I was growing up, honey bees provided me with a sense of well-being and stability, and, again, the opportunity to spend time with my father. But although beekeeping was clearly in my bloodline, it would be decades before I would consider it to be anything more than a hobby. Or, better put, until I had run out of other things that I could do to earn a living. I have been either fired or strongly encouraged to move on from almost every other job I have ever had over the past thirty years, which makes for a rather unimpressive résumé, I admit.

Still, in the early 2000s, I was motoring along as a full-time tenured associate professor for the state of Connecticut. I taught a few different courses, depending on the semester, including the basic English and composition courses taken by nearly all college freshmen in the U.S. education system, as well as English as a second language. Despite the routine of the job, I loved teaching, being in the classroom, the excitement of sharing and imparting knowledge. And it never escaped my mind that I was the first Coté in our clutch to have earned a university degree, which made my parents proud—prouder still when I went on to teach at university level. I was also among the first generation on my father's side whose mother tongue is English, so I held my foreign students in high esteem and admired their pluck to start a new life under the yoke of a new language. But the monotony of departmental meetings, community meetings, office hours, and all of the mundane and mind-numbing obligations that went with the territory in academia started wearing on me.

I did well at this institution. I had positive student reviews, was faculty adviser to the honor society, taught self-defense courses, organized blood drives, and in general was part of the community. But I was becoming more and more disheartened with my professional situation in some very real ways, and despite the melodrama of the phrasing, the honey bee and her positive influences saved me from being stuck in a career that, while noble and worthy, had ceased to be fulfilling for me. I could have easily remained at a job that I grew to dislike in order to take advantage of the steady income and benefits and promise of a pension; a state job is like a lobster pot—not so difficult to get into but very difficult to get out of.

It was the allure of the bees that wrenched me out of it.

I had long ago graduated from helping my father with his colonies to producing honey from my own beehives. I sold some of that honey from a farm stand at my house and even supplied a small quantity to a few grocery stores in the area. My brother, Michael—also a beekeeper, as well as a police officer—and I sometimes sold at the local farmers' markets together, and we always assisted my father and helped each other during spring buildup and at harvest time, which lent a nice familial vibe to what we did.

While I was teaching, I received some unsolicited press coverage for my extracurricular beekeeping. This wasn't perceived as a good thing in the eyes of my administration. At one point the president of the college called me into her office to suggest that I keep my beekeeping life and my academic life separate. She was referring to an article about honey bees she'd read in *The New York Times* that I was interviewed for, in which I'd mentioned the name of the college. As far as I could tell, this was the only time our small college was ever noted by the Gray Lady, and the school's PR person had already congratulated me on the coup. But the president clearly didn't agree.

On the one hand, I couldn't do much about stories of my vocation and my avocation intertwining; my position as a professor at a state institution was public record, and if a reporter writing about my beekeeping wanted to mention at which college I taught, it was easy enough to get the information. On the other hand, I absolutely bore some—most, probably—of the responsibility for the ultimate dissolution of my relationship with the college. And as the bees continued to lull me away from my responsibilities at the college, dissolve it did.

One day, I called in sick but was subsequently spotted blathering on about honey bees on a live television show. The next year I was AWOL at work when I was noted, again in *The New York Times,* as being present at a bee-related fiasco in a piece that mercifully made no mention of my place of employment. I was clearly permitting bees to encroach into my professional life in a way that demonstrated that perhaps I ought to give in to their sweet pleasures and forgo another year of paper cuts and windowless classrooms.

Having been a professor since my late twenties, I considered being an academic as part of my identity. I took my title and the position seriously, dressed the part in a suit and tie—though not a tweed jacket with leather elbow patches—and truly tried my best to do well. This, though my own academic history and ascension into academia was unconventional, to say the least. I attended Brien McMahon High School in Norwalk, but quit after my freshman year. Despite having no high school diploma, I managed to start classes at Norwalk Community College a few years later simply because no one in admissions picked up on the fact that I had not finished high school. (Years later I taught at the high school from which I'd never graduated, and taught at the college that had admitted me as a student without a diploma.) Then, at the age of twenty-two, I taught tenth grade composition at the American School, Colegio Americano, in Guayaquil, Ecuador. That lasted several months, until the director, who was aware of my certification shortcomings when he hired me, found a replacement teacher who actually had the appropriate four-year college degree that I had not yet earned.

But I enjoyed being in the classroom, and in the end I bounced around, attending five colleges and universities in three countries before finally cobbling together a bachelor's degree at the age of twenty-six. No one was more surprised than I was when I proceeded to procure my master's the following year. I had doubled up on classes and worked full-time as a sort of house mother at a private Japanese boarding school called Keio Academy in Purchase, New York, and as a ballroom dance instructor at an Arthur Murray in Darien, Connecticut, where my job was fundamentally to drag elderly women around the floor and venture not to drop them.

Then, in the late 1990s, just after getting my master's in education from Manhattanville College, I was awarded a grant to teach at a newly formed English language department in a small university in Mostar, Bosnia and Herzegovina, through George Soros's Open Society Foundations. Following that I landed a job teaching for Greenwich Public Schools, where, once again, I did not shine my brightest. I loved the community and my colleagues, and certainly the students. Still, after two years I knew for certain that I was not cut out for working with middle school kids. The administration did not try to dissuade me when I left that gig to be a Fulbright professor teaching master's students in eastern Europe just prior to taking on the full-time position with the state of Connecticut.

All this to say that even though my own education had been erratic, I took all of my life experiences and tried to apply them in the classroom. I poured every bit of my heart and soul into teaching, be it an English language club at Ritsumeikan University in Kyoto, literature courses at Univerzitet Džemal Bijedić in Mostar, pedagogy courses at Universitatea de Stat "Alecu Russo" in Bălți, or English as a second language classes at Norwalk Community College and Housatonic Community College, both in Connecticut.

But, ultimately, bees won me over. I traded in my tie for a hive tool, my suit jacket for a bee veil, my briefcase for a smoker.

Whatever misgivings I had, once I made the decision to leave academics it felt totally right, and I never looked back. I spoke to my union representative. I signed exit paperwork with human resources and, excited and terrified, left my cushy, tenured state job. I did not even pack up my office. I left all of my books and files behind, put my keys on the desk, said a few goodbyes, and walked out of the ivory tower in the direction of my apiary.

It was frightening in some ways, exhilarating in others, to voluntarily walk away from such a comfortable situation—to leave health insurance, a pension, and a steady paycheck behind. And, truth be told, there couldn't be many better jobs: We worked no more than four days a week, a maximum of six hours per day if that, and generally just thirty weeks per year. It was the best-paying full-time job with part-time hours ever invented. But I dove headfirst into beekeeping and trusted that the ladies in the hive would save me from a terrible fate. My trust was well placed.

After spending a few years tooling around Connecticut, taking care of my own beehives, and participating in a few farmers' markets in the Constitution State, I moved my base of operations to New York City. I started by joining one market, which blossomed into a few, where I met many clients who requested beekeeping services.

Now I'm often asked to install beehives, as it has become popular in some circles to adorn the rooftops of upscale restaurants, hotels, and businesses with beehives. The flagship Brooks Brothers store on Madison Avenue and Forty-fourth Street in Manhattan now has four beehives on the roof, one for each of the original Brooks brothers. (Claudio and Debra Del Vecchio, the owners of Brooks Brothers, have been good to me. When I went off to Italy to give a TEDx Talk, they kindly insisted on providing me with a new outfit.) I also installed ten beehives at Green-Wood Cemetery in Brooklyn, where the remains of the Brooks brothers can be found. I manage several others in nearby Industry City in Sunset Park. My single largest partnership is with the Durst Organization, which has dozens of beehives on several of its properties in the Big Apple. I've given talks on the importance of pollinators (like honey bees) to our fragile ecosystem at Durst buildings in tandem with Helena Durst, who's a former board member of Just Food, one of the nonprofits that helped push for the legalization of honey bees in New York City. Then there are the InterContinental New York Barclay, where I work with Chef Peter Betz, and the New York Hilton Midtown, where Zac Efron and I filmed a Netflix special about sustainable food, and many other places with great views and high profiles that now have apiaries that I maintain. I love the fact that my office is no office at all, but rather rooftops with billion-dollar views.

Keeping bees in a densely populated urban metropolis such as New York City may seem like an unlikely enterprise, but in fact urban beekeeping has been around for millennia. It was practiced three thousand years ago in Tel Rehov in Israel by ancient Egyptians and Greeks, and has been a part of New York City for centuries. The orchards once in the Lower East Side—around what is now Orchard Street—were at one time flush with bees. Just after the Civil War ended, *Harper's Bazaar* ran a story about a woman who managed forty beehives atop a building on Broadway and sold the honey to make a tidy profit. In 1881, on the roof of 14 Park Place, the proprietors of *Bee-Keeper's Magazine* kept a sizable apiary, and produced queens for sale that were shipped as far away as "Chile, New Zealand and the Sandwich Islands." More than a century ago New York City orphanages and hospitals maintained rooftop beehives as a means of providing sweets for their respective kitchens. The history of beekeeping in New York City could fill another book.

In 1905, when the Parks Department of the Bronx advertised a city apiarist position that paid $1,200 per annum, the most qualified candidate, who had scored 97 percent on the civil examination, was a former schoolteacher, Emma Haggerty. She, however, was blocked from accepting the position since as a woman she was not permitted to wear pants, which are generally preferred when working with bees, and part of the uniform for the job. The authorities seemed "to believe that one of the qualifications for political beekeeping is the right to vote and wear trousers." The *Brooklyn Eagle* protested, "A woman who is willing to take the chance of getting stung at swarming time ought to be a pretty good candidate."

There hasn't been a lull in New York City beekeeping activity for hundreds of years. Honey bees have been in New York City for quite some time, buzzing about and minding their own beeswax.

For eleven consecutive recent years, though, beekeeping in New York City was a covert affair, complete with underground networks of enthusiasts and contraband honey exchanges. In the decades prior to 1999, beekeepers in the city generally kept a low profile, but no one was worried about breaking the law, because there was no law to break. But under the Giuliani administration in 1999, beekeeping, specifically "the harboring of honey bees," became illegal. So, in 1999, when the last of the checkered cabs disappeared from the New York City landscape, honey bees became by decree *insectus non gratus* in the Big Apple.

It was not again a legitimate undertaking until the New York City Beekeepers Association, Just Food, and other like-minded groups got together and petitioned the city to allow the practice in the five boroughs. After all, at that point urban beekeeping was permitted in Boise, Idaho. Idaho! No disrespect at all intended to the great city of Boise, which has much to be proud of, what with its flat-track roller derby league, the Treasure Valley Rollergirls. But there is no reason for New York City to play second fiddle to the eightieth largest metropolitan area in the country. New Yorkers like to lead the pack. I found it mildly humiliating that my fellow city residents and I could not maintain beehives while Boise could enjoy both potatoes *and* all the fresh honey they could get their bees to muster. Thank goodness that since 2010 the Department of Health and Mental Hygiene—which issues dog licenses, handles tuberculosis control, and oversees issues relating to honey bees, now once again bestows their blessings upon the fine (and less fine) beekeepers of Gotham City in ruling urban beekeeping permissible. It is also worth noting that the name DOHMH was not chosen specifically with beekeepers in mind, though it does feel appropriate for the many beekeepers with emotional issues.

During the last weeks of February, most human beekeeping activity is as dormant as the honey bees are. Except for those rare balmy days when we're checking on our hives, we beekeepers are holed up in our homes, huddled against the elements, planning and plotting for the season ahead: how to improve on a hive configuration, how to maximize yield, how to find new and better locations for honey bees, how to conspire at once with and against Mother Nature to make the best of the bees' and our joint labor.

The New York City Beekeepers Association—of which I am the de facto head, having founded it in 2007 when I was feeling a need for some unity and camaraderie among urban beekeepers—usually holds introductory beekeeping classes in February and March, and teaches theoretical beekeeping to lead up to the practical beekeeping that starts in April. Classes are usually capped at fifty people in order to handle the questions and needs of the students. Additionally, we have an apprenticeship program that takes a dozen people per year; we hold meetings with speakers who are specialists in the field; we have group harvests; and we do the sorts of things that beekeeping groups all around the world do. The bees are the glue (propolis, perhaps) that binds us together.

* * *

There are two castes of female honey bees, the queen and the worker, and one male caste, the drone. Of the tens of thousands of honey bees in a colony, several hundred at most are drones— always a small minority. Drones are as few as they are misunderstood, yet they are as vital to the colony as any of the other players. The drones are the unsung athletes of the hive. They spend many afternoons in active mating flight, making up to half a dozen trips to a drone congregation area in search of unmated queens. When not pursuing their intense sexual search, they take a good amount of rest and feed well on pollen and honey to keep their strength up. Still, these male bees don't work, can't sting, and, though most never realize their amorous ambitions, would seem to live only to copulate with a queen. They have something of a tarnished reputation among humans. This is not out of line with most males of any species. It's not all the drone's fault. He grew up without a father. In fact, he has no father. It is not that his father went out for cigarettes one day and never returned—he actually has no father. The drone is hatched from an unfertilized egg, and has only the single (haploid) set of chromosomes from the mother, hence no father. His mother, the queen, has two sets of chromosomes, so she has both a mother and a father, giving the fatherless drone grandparents. Drones have a fascinating and unique anatomy. Imagine a penis that resides, turned inside out, within the abdomen, emerging for only one explosive ejaculation per lifetime. One. That's it. And afterward, the drone dies, having ejaculated with such force that it killed him. That's what makes the drone so special. Explosive balls.

Of the triumvirate within the hive, most people think of the queen and the worker bees before they think of the lowly drone, if indeed they think of him at all. The very word has taken on a pejorative meaning; it is not a compliment to be told that one "drones on." Yet there is much perspective to be gained about him. His is a life all about service and sacrifice. Sometimes, like a well-meaning guest, the drone tries to earn his keep. He might be called upon to keep the brood warm, which he does by planting his bulky, largely useless self atop the babies in their cells. He assists in circulating airflow within the hive by flapping his strong wings—this to reduce dampness and to keep fresh air in the chambers. He may even attempt some saber rattling in defense of the hive, by making as much noise as he can with those same wings when enemies appear—although, without a stinger, this amounts to a boisterous show of smoke and mirrors derived from his loud hum and sizable stature. He can't really defend the hive. He's a loudmouth with nothing to back up his bravado. Like lots of men.

It is easier to explain what the drone cannot do rather than what little he can. The drone, who is born without the abilities and physical attributes of his sister the worker, is unable to gather nectar, pollen, or water for the colony. Since the drone is unable to sting or to steal honey, he is permitted to enter any hive, not only his own, and this only if the guard bees allow his reentry. During his meanderings from hive to hive, he might well spread disease from one abode to the next. This does not enhance his reputation among his co-workers or keepers. In fact, he is often kicked out of the hive once spring and summer have lapsed into autumn, and will die of starvation or exposure—that is, if he was not barred from the hive after one of his trips to the drone congregation area. In any event, by then, his services are no longer needed. He is something like a relative who visits, eats, borrows the car and a little pocket money, and then remains weeks beyond any reasonable gauge of decorum. Eventually, he is given the boot.

So there he is, the poor, misunderstood drone, an awkward oaf at the cool kids' table. His primary purpose is to fertilize a queen, should such a need arise. So, in fact, he *is* the party when that happens. Though the drone has no stinger, he does have an endophallus. Like the worker bee's barbed stinger—though not the same in terms of evolution or function—the drone's penis is barbed. In both cases, it is designed so that once inserted, it stays where it is put, because it is fully inverted into the body of the queen, breaking off at the boundary to the outside after entry. Physically, he has a stout body and large eyes. He needs that vision in order to spot a randy queen when she is seeking a series of mates for a tumble in the clouds. As he is large and full-bodied, he stands out from the female workers, and is often mistaken for the queen by novice beekeepers. He is usually shorter in length than the queen bee, his stout midsection a direct contrast to her canalboat-shaped, tapered abdomen. Despite his hefty size—he reminds me of the AMC Pacer my mother drove when I was a child in the seventies—the drone can and must fly very fast in order to catch that vixen of a queen when she is courting.

This once-in-a-lifetime feat of athleticism occurs in the drone-congregating area, a watering hole in the sky where thousands of drones from various colonies congregate and wait for a receptive queen to enter the zone. Think, if you will, of a bar whose clientele is exclusively male, yet day after day the men hang around hoping for a woman to walk in. In the bee world, when the virgin queen departs from the hive on her mating flight, she enters the drone-congregating area. The drones detect her via thousands of receptor plates on their antennae, skipping the drinks and the small talk. The fastest and the strongest catch up to the queen and mate with her. But it is suicide by sex. When the drone's endophallus enters the queen, his abdomen contracts and his muscles tighten. This restricts airflow and the drone soon dies.

His big eyes blazing and his manhood engorged, if the drone flies fast enough to beat out the majority of his contemporaries, he mates with the queen. It's better to be the first to mate with the queen, simply because any drone that comes after must first remove the previous drone's endophallus prior to inserting his own. In fact, the drone comes equipped with a small hook on the end of his erect penis (or maybe it is on one of his legs, my sources vary), and with that he grips and removes the genitals of his predecessor prior to inserting his own pride. An average of fifteen of the strongest and fastest drones catch and mount the queen, and mate with her—which all takes place while the respective parties are flying upward toward the stratosphere.

Upon completion of the act, which is under five seconds*

---

* So that's two to five seconds of lovemaking for a six-week lifetime—which seems brief—but to be fair, if a human man lives to be sixty years old, and a drone lives six weeks but copulates for even two seconds, that figures to be the equivalent of more than half an hour for the human. Not so shabby—especially considering it is his first time. And theoretically, unlike his bug-eyed brother, the human might repeat the act.

(but five seconds of a six-week lifetime) per drone, with the entire event over in under a few minutes in most cases, each suitor has his penis torn out of their abdomen and they fall dying to earth, presumably with mixed emotions. They may or may not leave their hearts literally as well as figuratively behind, as their heart is located in the abdomen, and their abdomens and their bits certainly tear away. Thus their manhood, such as it is, is left within her. That spring day, the queen then returns to the hive and usually never leaves or mates again, unless she leaves the hive to swarm. And the next day, somewhere just outside the hive, piled irregularly, there is a heap of broken-off drone phallus tips, or remaining pieces of phalli, that were removed from the inside of the queen bee by faithful worker bees, who surely have a difficult time explaining that task on their résumé.

But in February, at least in the Northeast, the idea of mating is still just a twinkle in the oversized eye of the drone. February hereabouts is cold, certainly too cold for mating and usually too cold for leaving the hive at all.* My father, who is now in his late seventies and still lifts weights several times per week and works minimum twelve hours per day, used to take care of beehives for a number of families and estates in Westchester County, New York, and Connecticut. One day in February he got a call from a woman he'd met at the gym. As mentioned, Norwalk is surrounded by well-to-do towns, and this woman had hired him to take care of beehives on her property. Among other endeavors, she had been a caterer, and around this time she'd started to produce a television show out

---

* In warmer climates, the opportunity for mating exists year-round.

of her home on Turkey Hill Road in Westport, Connecticut. She had a nice garden, scores of rare-breed chickens like Ayam Cemani and Araucana, which laid more eggs than a small army could consume, and an extensive collection of antique stoves kept in a barn. Her name was Martha Stewart, and she has come to be well known for her skills in the kitchen and in the household. In addition to maintaining her hives, my father attended parties and other functions as her guest.

On the February day that she called him, it had been snowing and she was worried that her bees were dead. "There are so many bodies out in front of the hives!" she told him.

"That's a good sign. It means that the bees inside are alive and throwing out the dead."

"Can you just come and check on them?"

Pause. Sigh. "Okay, Martha."

So Norm got into his pickup truck and drove to Westport in the snow. He checked the beehives with a stethoscope and found them to be alive and well, as expected. He was ready to leave when Martha, in her pajamas, called for him to come into the kitchen. She asked him if he'd like some coffee. "Or even a cappuccino?" she enticed further.

Norm looked past her to see the impressive machine in which Martha would make the beverage. "It looked like a spaceship, but with more knobs and dials," he told me later. He started to step into the house but was quickly told to take off his shoes before entering the kitchen. Soon, he and Martha were sitting and sipping their hot, foamy cappuccinos and talking about bees. "I should have stolen a few fancy eggs from the coop on my way out for the extra time," he said, laughing.

Martha was a stickler as a client. For most clients, he would usually remove full honey supers—the boxes filled with surplus honey meant to harvest—and take them to his place to extract the honey. But Martha insisted that everything be done on her premises to ensure that it was not commingled with any other honey.

My father oftentimes took me with him to his clients' homes to assist him, and to learn from him. I never spoke too much with Martha during any of these visits; our acquaintanceship came much later, when she would visit me at the Union Square Greenmarket and later when I appeared on her show. My father observed, "She *is* nice, but she likes to talk. They all do. It's nice to talk about bees, but you can burn your whole workday talking about bees and not getting any bee work done. So wherever you go, it's better to get in and out without being seen."

In addition to Martha Stewart, my father had quite a few clients and a long waiting list of hopefuls who wished to obtain his services. This made him selective. He preferred clients who left him alone, let him manage the bees, and didn't bother him while he was working. Those who were omnipresent or just too chatty would find, the following year, he sadly could not fit them into his schedule.

He just liked working with the bees, and did not want to spend hours every day talking to people. His hearing is poor. "They are all very nice and interested, but I cannot hear them half of the time, and all I can think about is how many more hives I need to check before it gets dark," he would say to me. All of his clients were from word of mouth. Included in his route were the well heeled—the homes of George Soros, the Rockefellers, and other prominent families with enough land on their estates that he could easily go for years without seeing anyone other than his fellow hired hands as he drove in and out of their guarded properties to tend to the beehives.

While Martha did chat with Norm from time to time, he really did like her, and so of course he continued to keep her on. She left Westport, Connecticut, behind and moved to a gorgeous piece of land in Katonah, New York, located in Westchester County. I often accompanied my dad to the estate. Once we passed the ever-present armed guards, we often first checked in and visited the ornate stables—much larger and better built than our own houses—where a team of impressive breeds were looked after by a crack team of Nepali horse trainers. The stable floors were heated, the lightbulbs, even high up, were dusted and wiped daily. It was a two-tiered, magnificent place. The fence that surrounded the paddock was made from petrified wood and had been brought down in pieces from a ranch in Canada. These were some high-living horses.

My father's favorite part of the exquisite property was the garden—and he rarely failed to wander past it without grabbing a tomato or something to munch on. "They taste better when you steal 'em," he told me with a smile, juice dripping down his chin. And then we would tend to the bees.

There were four beehives that Martha wanted painted "Katonah colors," which meant her paint brand at the time. No problem, since they really were lovely colors. We could never fault Martha for style. When my father appeared on her show many moons ago, each small silver spoon used to gently ladle small amounts of honey was worth several thousand dollars. Martha didn't mess around. She had a separate washer and dryer for towels that were used exclusively for her cats. We are all aware that Martha has peculiarities or eccentricities. But these are hard earned. She built an empire out of arranging flowers and baking pies. I have always admired her. Even after she fired my father.

Actually Martha fired my father three times, but it only stuck once (the last time). They had a funny relationship. The two of them exchanged hundreds of emails via a secret email address—or at least a non-public one—that Martha uses to communicate with people directly. Martha was very candid with Norm. He enjoyed their exchanges and friendship, even though sometimes he grumbled. He never cared about her fame and fortune. Even after it was clear that Martha was becoming a worldwide icon, to him she was just a woman he'd met at the gym in Connecticut a couple of decades prior. He started taking care of her bees when she learned that was what he did, after she'd spotted his smoker, the Tin-Man's-head-looking device in the back of his little pickup truck, still smoldering. They became almost chummy. He attended a 1950s-themed party at her house, where my mother dressed in a poodle skirt and my dad wore jeans, a white T-shirt, a leather jacket, and generous amounts of Brylcreem in his hair. In a quaint way, he and Martha were pals.

The first time she dismissed him was when she was upset that one of her beehives had been knocked over during a hurricane. "Martha, that storm knocked down a dozen trees on your property, too. And the bees are fine." For whatever reason, it stuck in her craw, and Norm was told his services were no longer needed. But two weeks later she started writing to him about the bees again as if nothing had transpired. So he kept going and didn't mention it. He never said it to me, but I know he was happy to be back. The second time she thought she would try having one of her other workers, a landscaper, take over. He claimed he had taken care of bee colonies back in his hometown of Cuernavaca, Mexico. We are always a bit suspect of this sort of claim—loads of people we meet tell us, "Oh, I took care of bees in my country," wherever that may be. Anyway, after one honey-free season and four dead beehives, Norm was reenlisted.

One day, during his third installment, Norm was casually talking with the property manager, and a few of the other workers were there—including the fellow who'd had a brief tenure as beekeeper. Norm and the manager were chatting about this and that, and as the help are wont to do, may have been jesting a bit about the lady in the "big house." Norm noticed that there were feeders on the beehives, but it was far too early in the year for that, and it was not good for the bees to have that cold liquid potentially dripping onto them while they were clustered together trying to stay warm and dry. The property manager was saying that Martha disagreed and had ordered the feeders on (this was probably not the case—it was more likely one of the workers who felt he knew best and did not want to back down—Martha is savvy enough to defer to experts when it matters). Norm explained that keeping the feeders on was just not advisable. He tried to soften the blow since he knew that probably whoever made the poor suggestion was there and Norm did not want him to lose face or have a confrontation. So he tried humor. "You tell Martha to just stick to baking the cookies and leave the beekeeping to me," he told the property manager. The fellows around laughed, but not the property manager. He just made a sour face. Norm realized he may have stepped in it. In fact he was sure that he had.

It is one thing to blow off some good-natured steam at the expense of the matriarch; it is quite another when she hears about it. And so Norm's time, and by default mine, at the estate came to an abrupt end. No more of Martha's garden tomatoes for the Coté clan.

MARCH

Jim said bees wouldn't sting idiots; but I didn't believe that, because I had tried them lots of times myself, and they wouldn't sting me.

—MARK TWAIN, *Huckleberry Finn*

Spring begins to spring. Skunk cabbages blossom. New Yorkers walk around with their coats unbuttoned. The chimes from the Delacorte Clock in Central Park start playing "Easter Parade." Nearby, the Shakespeare Garden shows off blooming forsythia, Arabis, daphne, witch hazel, crocus, and snowdrops, and a few tulips and daffodils peep up from the cold ground. Later in the month, if I am lucky, I have time to see the cherry trees bloom just east of Strawberry Fields. Soon my bees shall feast upon the bounty within Central Park and beyond. This banquet will culminate in a unique and light honey found nowhere else in the world, thanks to the blend of nectars from a grouping of nonindigenous flowers all concentrated in the park's eight-hundred-plus-acre playground. Once harvested, the honey sells as fast as I can bottle it to tourists visiting the city and to New Yorkers wishing to consume the local pollens in the raw honey to stave off seasonal allergies.

For me, March always brings a sense of excitement, as there is so much to look forward to with the promise of a new beekeeping year. This is tempered with a mild sense of foreboding, as I know the huge amount of work that lies ahead. For the bees themselves, it is a risky time. Even if they survived their sequestered state within the hive with nothing to feast upon other than what they gathered the previous year, starvation is still a threat. The bees cannot safely leave their home to look for nectar if the external temperatures are not at least 55 degrees Fahrenheit, so a colony lucky and strong enough to have survived the long winter may be in danger of starvation just because it is cold outside and they cannot leave the house. Or the temperatures may rise just enough that they can fly around in search of food, but with relatively few flowers having blossomed, there may be little to none yet available to them. If there are reasonable reserves of honey, there should be a steady increase in egg laying by the queen. If so, there will soon be many more mouths to feed, and if the weather doesn't cooperate, the winter supplies may be gobbled up without an opportunity for the bees to gather more.

Additionally, the bees will have likely broken their winter cluster by now, but if a cold snap descends on the area, it could devastate even an otherwise healthy and well-stocked colony. The bees might sense the temperature dropping and rush to regroup themselves, inadvertently creating two clusters, neither of which will be able to sustain the necessary temperature. In this case, the entire colony would perish.

So March is exciting, yes. But it's risky. That's why at this time of year, a beekeeper will often feed fondant or sugar water to the bees until they can start to gather their own nectar. This in no way impacts the honey that will be harvested later in the year. While there are those who believe that under no circumstances should honey bees ever be fed sugar water, I equate feeding honey bees emergency sugar in the off-season to giving a child an unhealthy snack like a bag of chips when a real meal is still a few hours away; it's a necessary evil that isn't so sinister.

As the daylight hours start to increase, and if the temperature is agreeable, a beekeeper might take a furtive look under the cover of the hive just to see if the girls rallied through the winter. Frames should not be exposed to the air if it is below 60 degrees Fahrenheit, and even more care should be taken if the winds are high. This is to protect the eggs, or the baby bees* who are just coming up in their world. But for the purposes of a quick pre-spring check, there is no need to remove frames and do a full inspection; a brief look to see if there is life will suffice. If yes, rejoice. If not, time to clean out the equipment and hope that there is no mold or wax moth damage, or that mice have not made the dead-out beehive their abode (and eaten the wax for good measure). On the odd warm day, the bees will take what is called a cleansing flight. Since honey bees, as a general rule, will not empty their

---

* That is, larvae, or immature bees.

bowels inside the hive—only her majesty the queen may plop her royal poop within the hive with reckless abandon, and that regal excreta is carried out joyfully by her attendants—they will hold their own bowels until the weather permits them to fly out and void them. This is even true in outer space.

Whereas astronauts may at times wear diapers—which NASA calls maximum absorbency garments—while on missions, honey bees will do their very best to hold their bowels until they can leave the beehive. In 1984, aboard the space shuttle *Challenger*, seven thousand intergalactic honey bees had to keep their three pairs of legs crossed until they got back to Earth. Conducting an experiment testing the impact of microgravity on honey bees demonstrated a few things. Honey bees can actually feel gravity and use that to hone their honeycomb-making skills. In space, the lack of gravity influenced the bees to build their honeycomb at odd angles and not to the perfect pitch of the Earth that they normally would. They also, initially, could not fly, and just walked around, but by the end of the week they had worked out how to fly in their new environs. Honey bees are quick studies. And lest one think that honey bees on the space shuttle is odd, they aren't by a long shot the most curious creatures taken aboard the various shuttle missions. Sea urchin sperm, worms, and jellyfish have also made the voyage into space.

Back on planet Earth, specifically when I was growing up in Connecticut, our wonderful neighbor, Mrs. Berger, a German woman and the mother of five boys, always washed her family's linens and hung them to dry on a line in the backyard that was fairly close to my father's beehives. Every springtime, she would assert, "Something must be coming off the trees and marking on my sheets!" in a German accent as thick as the butter she would spread on my toast. For decades, this lovely woman has believed that the trees dividing our yards shed something that caused brownish yellow stains on her crisply washed and hung white sheets. No one has had the heart, or courage, since she is as tough as she is kind, to tell her that the streaks are, in fact, bee *Schiesse* being released from above by our bees each spring. If you are reading this, Mrs. Berger, thank you for letting us use your pool for all those years, and for all of the sodas you snuck me when I was little. And on behalf of my father and our bees, I am very sorry about the sheets.

All worker bees are females, and at this time of year especially, they live up to their name: They *are workers*. And they are, all of them, sexually underdeveloped queens. Had they been afforded a nutritious diet of royal jelly for their entire incubation from eggs to pupae to larvae to emergence from their cells—a twenty-one-day journey from being laid to biting and scraping their way out of their cells—then they would have developed ovaries and become queens. So every worker bee—that is, every female bee—could have been a queen had she been provided a finer repast. But, as it plays out, the nurse bees (worker bees, of course), whose responsibility it is to decide which fertilized eggs will become queens, make those decisions on an as-needed basis. A kitchen cannot function with all chefs, and a beehive cannot run without the vast majority of its population being worker bees.

A worker bee generally lives about six weeks, with roughly half of that time spent on indoor tasks and the other half spent on foraging for various items needed by the colony. From the moment the worker bee chews her way out of her cell, she commences a life of toil. During the first two weeks or so, she will work in part as sanitation engineer, cleaning out cells and making the area impeccable for the next round of eggs.

Worker bees also function as undertakers, dragging out the dead from the beehive. Sometimes the dead are not even all the way dead. Watching undertaker bees drag out placid and sick or injured-but-still-living bees puts me in mind of the scene in *Monty Python and the Holy Grail* where a man is pushing a cart during the Dark Ages collecting bodies, one of which protests "I'm not dead. I'm feeling better." Then he's clobbered and heaped on the cart with the other corpses. But whether it involves the dead or deadweight, a beehive or a plague-infested village in the Middle Ages, housecleaning must be done. And as honey bees are immaculate creatures, the corpses will be removed as far as possible from the hive to avoid disease.

These young worker bees are also caterers and waitstaff to the baby drones, since the young men cannot yet feed themselves. And, still in the first fortnight, some of the young workers are charged with taking care of the queen. This includes her grooming, feeding, and guarding. So they are beauticians as well. They also disperse the queen mandibular pheromone (QMP) throughout the hive. Since the honey bee world relies heavily upon scent, the spreading of this scent is essential to a healthy and functioning community. When the Royal Standard flag is flown at Buckingham Palace, it is an indication that the queen is in residence. The QMP that is spread by young worker bees is something of an olfactory equivalent. It lets the general population know that their queen is stalwart and that all is well in their world. So in this sense the worker bees are disseminators of information.

The queen bee eats only royal jelly, and royal jelly is secreted from the head of the nurse bee via the glands in her hypopharynx. So add caterer and chef to the job description. Lest anyone think that only bees eat strange things produced from their own bodies, in the same month and year that bee-keeping was legalized in New York City, there was a well-known chef in Manhattan who made and sold cheese made from his wife's breast milk. "I prepared a little canapé of breast-milk cheese with figs and Hungarian pepper," he told the *New York Post*. Chew on that.

Sometimes the bees themselves are the meal. In Japan, there is a region in a remote mountainous highland valley area called Kamikōchi (上高地), where one of the more eso-teric culinary specialties is honey bee larvae. It is paired with horsemeat. Emperor Showa, formerly known as Hirohito, was said to be a big fan of eating honey bee larvae.

If they escape the chopsticks in that highland valley, dur-ing the third week the worker bees become warehouse work-ers and professional organizers, taking the pollen that their elder sisters bring back to the hive and storing it in the honey-comb. They are also chemists, as they mix the pollen with a bit of honey to create what we humans call bee bread, a concoc-tion used to feed the babies. This third week they also become HVAC technicians, as they use their double set of wings to circulate air within the hive in order to keep the temperature down.

Add military adviser, bodyguard, and security guard to the list of jobs for the third week, since from about the eighteenth until the twenty-first day of their lives worker bees keep a wary eye out for any shenanigans from intruders. Or in their case, five eyes, each worker being equipped with three ocelli eyes that are simple and read light intensity, as well as two larger compound eyes with nearly seven thousand facets apiece built for detecting movement. These guard bees will maintain their vigilance at the front door of the hive with increased numbers in late summer or early fall, for when winter is coming, the chances of being attacked and robbed increases. It is no game of drones.

Still in the third week, workers become structural engineers, craftswomen, architects, and day and night laborers when they begin to build honeycomb. Honeycomb is built from wax that the bees produce from their own bodies—another reason to give bees sugar water in the early part of the year, as it stimulates their ability to produce wax. Bees convert sugar from honey or syrup into wax, which they push out of pores in their abdomens in small flakes. They produce this wax in a similar fashion to how humans produce ear wax. The workers then chew on the wax to render it pliable and use it for the construction of the honeycomb, which is their home; to seal cured honey; and for other construction projects.

Gunga Din had nothing on honey bees. Like all living creatures, honey bees need water to survive. The water carriers work in tandem with the fanning bees in order to keep the temperature of the hive down on hot days. In the 1930s, trains in India may have taken a cue from honey bees when they would position fans behind huge blocks of ice in attempts to lower the temperature of the carriages. And speaking of honey bees and India, there is a hill station in the southern state of Kerala called Munnar. I have been there and stood in wonder at one particular tree on Mattupetty Road where there hang about three dozen feral beehives. The comb is exposed, rather than hidden in a safe cavity as is usually the case with honey bees, and the colonies somehow live more or less in harmony. Oddly, the many surrounding trees are untouched by the bees. For many years they have adorned just this one tree.

Surely the best-known job of the worker bees is that of forager, which is their role for the last three weeks of their lives. Essentially, foragers fly out from the hive, beginning with short expeditions to orient themselves. In time they alternately gather water, resin (to make propolis, which is a bee glue, and used to seal cracks in the hive), nectar (to transform into honey), and pollen (their source of protein). Though they are able to sting, workers will do so only to protect themselves or their colony. Their stinger is a modified ovipositor—though worker bees generally do not lay eggs. Though queens and workers evolved stingers from an ovipositor, both can lay eggs via the sting chamber. But not to worry; honey bees found on flowers are harmless and have no interest in stinging anyone or anything. They just want to do their job.

Just as March ushers in the spring and much of nature begins to stir, we beekeepers become more active as well. This is the month when the New York City Beekeepers Association finishes up its basic beekeeping course. Since 2007, the NYCBA has rented out a hall and taught courses to beginner and wannabe beekeepers over several weekends during the winter months.

We started by renting out small rooms at a YMCA in

Chinatown in lower Manhattan, and then moved into a basement conference room of a high-rise hotel where I maintained
beehives on the roof of the seventy-second floor in what is the
highest apiary in the world—though not the most productive,
as the poor foragers have quite a time making it the one-third
of a mile to the top of that building, fully laden with nectar
and pollen, and in sometimes brutal winds. More recently we
have been using space at the New York Institute of Technology on Broadway and Sixty-first, and it is by far the best space,
equipped with large-screen monitors to display the hundreds
of videos and photographs that accompany our courses. What
started off as a dozen or so students in 2007 rapidly grew to
what now has to be capped according to room capacity. The
classes have been full since 2010, when beekeeping became
legal in the city. The week that the city of New York was to
hold public hearings on the issue of legalizing beekeeping, *The
New York Times* published an article about it, which read in
part:

> New York City is among the few jurisdictions in the coun-
> try that deem beekeeping illegal, lumping the honey bee
> together with hyenas, tarantulas, cobras, dingoes and other
> animals considered too dangerous or venomous for city
> life. But the honey bee's bad rap—and the days of urban
> beekeepers being outlaws—may soon be over. . . .
>
> The Department of Health and Mental Hygiene's
> board will take up the issue of amending the health
> code to allow residents to keep hives of *Apis mellifera*, the
> common, nonaggressive honey bee. Health department
> officials said the change was being considered after re-

search showed that the reports of bee stings in the city were minimal and that honey bees did not pose a public health threat.

The officials were also prodded by beekeepers who, in a petition and at a public hearing last month, argued that their hives promoted sustainable agriculture in the city.

This was, of course, after about two years of petitioning, meetings with city officials, conversations with groups like Slow Food and Just Food, and holding powwows of our own with NYCBA members. I worked a great deal with now-retired Nancy Clark, then Assistant Commissioner for Environmental Disease Protection at the Department of Health and Mental Hygiene (DOHMH). She openly wanted to help us in our endeavor but wanted to do so safely—a shared goal—and in a way that would limit liability for the city and not put more responsibility on the shoulders of an already overburdened DOHMH. In the end, the results were bittersweet. We won the battle to legalize harboring beehives, but with virtually no regulation or enforcement, it has led to some questionable beekeeping practices that are technically permissible. But it has worked out well overall.

In any event, the day that there was to be a city council public hearing and vote on whether to lift the restriction on harboring honey bees, the hearing room was packed with a standing-room-only crowd. But the mass turnout was not solely for the humble honey bee; there was a strong equestrian element in the room. Though horses and bees do not get along, apiarists and hippophiles can be quite chummy. As it turned out, most of the meeting's attendees were horse lovers who were there to engage in what became an astonishingly vicious—and vociferous—battle of words between those who wished to end the practice of allowing horse-drawn carriages in and around Central Park and those who opposed any change to the regulations. Emotions ran so high one would have thought old nags were being turned into glue right there in front of us, and that the event was being catered by a group from Kamikōchi serving their regional specialty. Amid the fracas, anyone who wished to speak was given five minutes to make their remarks. There were two stenographers, one to take down testimony regarding the bees and one regarding the horses. During my five minutes, dressed like a big boy in a suit and tie and seated on the stage before the panel and the audience, I mingled my comments on bees and horses just enough to watch the two stenographers eye each other and laugh in confusion.

In retrospect, the heated exchanges and heightened emotions in the room may have drawn angst away from the relatively serene debate on beekeeping. Though "debate" may not be the right word—not one person spoke against the idea of making beekeeping legal, and the measure passed unanimously. The health code now reads in part:

Section 161.01(b)(12) . . . requires beekeepers in New York City to "adhere to appropriate beekeeping practices including maintaining bee colonies in movable-frame hives that are kept in sound and usable condition; providing a constant and adequate water source; locating hives so that the movement of bees does not become an animal nuisance, as defined in § 161.02 of this Article; and shall be able to respond immediately to control bee swarms and to remediate nuisance conditions." Section 161.02 defines a beekeeping nuisance to "mean conditions that include, but not be limited to, aggressive or objectionable bee behaviors, hive placement or bee movement that interferes with pedestrian traffic or persons residing on or adjacent to the hive premises; and overcrowded, deceased or abandoned hives."

Of course, things don't always work out as would be ideal. Bees do indeed swarm, people are inattentive, and there are consequences to poor management or just unlucky situations. DOHMH can levy fines for infractions that measure in the thousands of dollars. Still, there are more pluses than minuses to keeping honey bees in an urban environment.

As for being safe, certainly it is safer to keep a box of bees on a balcony than to try to cross Queens Boulevard on foot, the thoroughfare nicknamed "the Boulevard of Death" due to the high mortality rate of pedestrians who unsuccessfully attempt crossings. And in New York City, the danger of being eliminated from the population via honey bee stings is far smaller than the danger from falling appliances; many more people have received their own angelic wings by falling victim to plummeting air conditioners when simply strolling down a city sidewalk. Lastly, at the time that the city's health department decided in its infinite wisdom to allow those of us who want to keep beehives to do so, First Lady Michelle Obama was busy setting up not only an organic garden but a beehive on the White House lawn (she also visited my beehives atop the Waldorf Astoria). It is safe to say that if the Secret Service allowed the two most heavily protected children in the country to frolic in close proximity to seventy-five thousand bees—Sasha and Malia were known to play in that area—the flying creatures can be assumed not to pose a serious threat.

The practice of high-ranking United States politicians keeping colonies of honey bees did not start with the Obamas. Way back in 1787, George Washington had a "bee house built on the grounds of Mount Vernon which he had inherited," according to the late great Eva Crane, quantum mathematician turned bee researcher.

Washington was not the only president to keep honey bees, or to at least have an interest in them. Our third president, Thomas Jefferson, was also fond of honey bees and honey. Jefferson's overseer, Edmund Bacon, was a beekeeper. Bacon wrote of how Jefferson came to visit his forty colonies of honey bees. Jefferson's interest was not in passing. He discussed honey bees in his *Notes on the State of Virginia*, saying in part, "The honey bee is not a native of our continent. . . . [Honey] bees have generally extended themselves into the country, a little in advance of the white settlers. The Indians therefore call them the white man's fly, and consider their approach as indicating the approach of the settlements of whites." Jefferson even owned a British book from the mid-1700s entitled *Collateral Bee-Boxes: Or, a New, Easy, and Advantageous Method of Managing Bees*. As of this writing, beehives are kept at Jefferson's former property, Monticello, and the honey from those hives has been used in the Virginia governor's mansion and even at the White House. As it happens, Jefferson appears on the two-dollar bill, which I keep stacks of at the farmers' market, and give customers change in the under-circulated note. The bill is attractive, unusual, and in most cases is a nice conversation piece. In almost all cases it puts a smile on the customer's face.

Centuries later, the Obamas, thanks to the first lady, kept a beehive on the White House lawn, near to where Woodrow Wilson used to have his sheep graze. When the Obamas left the White House, The Trumps elected not to keep the beehive, but that same year an apiary was established at the vice president's residence, and Karen Pence, wife to Vice President Mike Pence, tends to them, just as she did back when she was first lady of Indiana. So honey bees have been consistently a bipartisan interest for centuries in the United States.

As for the debate about horses and horse-drawn carriages? The winds of change did not carry the day on that issue, and the law remained stable.

I was traveling in those weeks of late March 2010, as beekeeping classes were finished but the season had not yet begun. I was on a plane bound for Guayaquil, Ecuador, to scope out a potential Bees Without Borders project. Upon boarding my flight I picked up a copy of *The New York Times,* knowing that there was supposed to be a piece about legalizing New York City beekeeping in it that I was interviewed for. I did not expect the story to be on the front page, nor was I prepared for the response to that. I soon learned that when an individual is quoted in a front-page *Times* story, quite a few journalists from around the world will try to locate and interview said individual.

Soon I was receiving email from virtually everywhere—the United Kingdom, Germany, Sweden, Russia, Australia,

China, and other spots around the globe. Reporters, bloggers, students, and other beekeeping associations flooded the club's email inbox with hundreds of requests for interviews, comments, and information. I ended up spending much of my time south of the equator commandeering the computer and telephone at a copy shop in order to field all of the requests from media regarding the legalization of beekeeping in New York City. It was eye-opening to see how powerful one article could be, and though I didn't yet realize it, it was a herald of things to come.

When I returned to the States later in the month, it was time to pick up a hundred fully established overwintered beehives I had purchased from a retiring beekeeper in rural Pennsylvania way out near the Ohio border. This, even though my friend Tammy Horn, a well-known beekeeping expert and author from Kentucky, had told me, "Never buy bees and equipment from a retiring beekeeper. Most of them have been retiring for the last fifteen years and don't make any repairs on anything during all that time."

Against better judgment, my father, some friends, and I drove two trucks and trailers the nine hours from our small farm in Connecticut and arrived on this fellow's (pronounced "feller" out there) farm in Otter Creek, Pennsylvania, just as the sun was setting. Many of the homes in the area were trailers on cinder blocks, including the home of the man from whom I was making the purchase. This feller had taken up beekeeping as a young teenager in the 1950s, when his parents were getting divorced and he felt lonely. His own son was somewhat interested in taking over but not to the extent that the father had built the business. So once cash was counted out on the kitchen table and recounted and stacked, our team screened in all the bees, loaded them onto the trucks and trailers, and turned around to head back home.

Of the hundred-strong colonies of bees we picked up—all in shabby equipment due to the beekeeper's long-impending retirement—I had pre-sold twenty of them to the Brooklyn Grange, the largest rooftop farm in the world. Founded in part by Ben Flanner in 2010, the Brooklyn Grange is known to urban rooftop farmers around the globe. I had known Ben, a Midwesterner, since 2008, when I first placed a couple of beehives atop a small rooftop farm on Eagle Street in Greenpoint, Brooklyn, that he cofounded and later left for greener, and much larger, rooftops.

I had been dealing with a fellow at the Grange named Chase Emmons. According to his business card, his title was "chief beekeeper." Chase has since parted ways with the Grange and has, I am told, left the state. Regarding these twenty beehives, Chase was meant to come and pick them up the same day they arrived back east at our Connecticut farm. He failed to do so and asked me to "hang on to them a little while." Requesting that someone hang on to twenty full, live, screened-in beehives is no small ask. Full beehives that are screened in will not stay alive too long, especially if there is a warm snap. So I unscreened the bees to allow them to take cleansing flights and gather what they needed.

They oriented to their new location, and we all waited for Chase to arrive to collect them. Finally, a week later, he arrived with two other beekeepers, both known to me. First was Stephanos Koullios, a Greek American whose father owned a furniture factory in Long Island City, Queens. Together he and I kept half a dozen beehives on the roof of the factory. And together we had filmed an episode of a BBC television series called *This Human Planet*, right there on that rooftop with views of the Queensboro Bridge. We had previously captured swarms together. He was a nice guy who did not seem to take too much too seriously. Also with Chase was a guy named Tim O'Neal, from Troy, Ohio. I had worked with Tim several times, too. Tim once told the *New York Post*, "Whenever I find a swarm in an odd location, I take it out with my bare hands." And no doubt he said it exactly in the drawl that one would expect from a guy from Troy, Ohio.

To simply shift twenty beehives from the front yard of a suburban neighborhood into the back of a rented U-Haul truck theoretically should have posed no problem for this triumvirate of veteran beekeepers, especially since it included a valiant fellow who boasted a propensity for picking up thousands of bees barehanded.

First, the men suited up in their gear. Except for Stephanos, who had neglected to bring his. Instead, he sat eating plantain chips (plantain = banana. Banana + bees = not ideal, as the astute reader will have learned). "I know it isn't the best idea, but these are really good." Stephanos smiled in his own defense and offered me a chip, which I declined with a grin, envisioning their immediate future. Tim was careful to suit up and tuck his big white tube socks into his tight-fitting skinny hipster jeans. This was not a bad idea—the tucking in the socks anyway. The skinny jeans may not ever be a good idea for most men. There was not even a muted trace of the bravado he had proclaimed to the *Post*, but at least he was gloved and veiled and ready for action. Chase was the most dramatic looking. He donned a full-body suit advertised as sting-proof that covered him from head to toe. It was vented and thick and had rolls similar to those of the Michelin Man. Or maybe more like the Pillsbury Dough Boy. In either case, the wearer was well protected from both bee stings and dignity.

I had been ready and willing to help the three of them, mostly because I wanted the hives off my property. As we had planned and agreed for the hives to be picked up a week prior, I had left them all quite close to the road. Since they had to be unscreened so the bees could fly freely, they'd created a real nuisance for pedestrians and cyclists with their proximity to the public way. This distressed me enough to want to volunteer my time to get rid of them in spite of the fact that it was a hugely busy time of the year for me. But the fellows showed up several hours late on the day they finally managed to get to me, and by the time they seemed fully ready for action I chose to abandon them to their own devices and head to bed, as I rise quite early every day.

I said good night and left my longtime friend Anna Veccia on the porch to supervise and ensure that they took their twenty (and only their twenty) beehives. I provided crank straps, screens, and a simple but necessary tool called a hive carrier in case they needed any of these things. The idea was to screen the bees into the hives, make certain that a crank strap held each beehive together, and then for one person to get on either side of each hive. Together the two would lift it up and onto the truck. If I had been 100 percent certain that they were actually going to appear as scheduled, I would have screened the hives in for them. But after several postponements I was concerned that they might flake out again. I would then be forced to unscreen and rescreen the hives yet again. I cared very much for the welfare of the bees, and I wasn't going to let them suffer.

The colonies were all within five to ten steps of their rented truck, with the gate only about a foot off the ground. In other words, it should have been a simple task. But apparently it wasn't for them. Anna reported that after attempting to screen the bees in and relocate one hive over the scant distance onto their truck, their shoddy screening work came loose. The bees, naturally, attacked. Thin skin on ankles and wrists was instantly penetrated, and increasingly shrill noises emitted from the men. In fact, I could hear the shrieking in my bedroom all the way at the back of the house, until I turned on the television to drown it out. I fell asleep to the sounds of a *Sopranos* rerun. The next day I woke to find all of the beehives still in the front yard. One or two had been shifted from their original spots, but all were present and accounted for. Later that day Anna telephoned and reported that "watching the three wildly prance around getting stung was better than being at a cabaret. Be sure to let me know if they try again, I'll invite some friends."

A day or two later, Ben Flanner, the Brooklyn Grange founder, came to see me at my market at Union Square. Chase and his cohorts had been trying to pick up those beehives for the Grange, and Ben still wanted them. I still wanted Ben to have them. In short order Ben and I came to an agreement. I would screen in and deliver the beehives to the Brooklyn Navy Yard, where the Brooklyn Grange is located and where this apiary was to be. So my father, Anna, and I screened in the colonies and loaded all twenty hives onto one of my trucks without mishap. Anna drove along with me to the Navy Yard.

The Brooklyn Navy Yard is located along the East River overlooking lower Manhattan in an area called Wallabout Basin. It was once home to the Canarsee tribe of Native Americans. When the Dutch took over the area in the 1600s, a farm was established on the tribal lands.

In 1781, a shipbuilder purchased part of the farm, and by 1801 the U.S. Navy had taken over the site. In 1831, Commodore Matthew Perry began his tenure at the Brooklyn Navy Yard. Commodore Perry, who commanded ships in the War of 1812 and the Mexican-American War, is perhaps best known for his role in the forced opening of Japan in the 1850s through so-called gunboat diplomacy.

Ships were built, repaired, and berthed at the Navy Yard all the way up until 1966, when the site was finally decommissioned. Now a portion of the two hundred or so acres is designated as a historical site, and the rest is zoned for mixed business use. Currently the space houses, among other things, a whiskey distillery, a manufacturer of military apparel, a facility that produces and sells kitchen countertops made from recycled glass—and, of course, a rooftop farm. And, as of this day, beehives.

Anna and I arrived at the yard where we were scheduled to meet the same three fellows who had attempted to pick up the hives in Connecticut. This time, they arrived only slightly late, and they were clearly taking no chances and ready to defend against a full onslaught. Tim not only had his full gear on over his skinny jeans and shirt, he had also brought a roll of duct tape. I watched, fascinated, as he proceeded to tape his pants closed at the ankles, tape his gloves around his wrists, and even tape his jacket around his waist. He must have used half a full roll of thick black tape to seal off any possible breach in his britches into which these tiny creatures could penetrate. Misfortune, sadly, disfavors the fragile. Halfway through removing the beehives from the truck, Tim's constricted britches split right up the middle, exposing his tighty-whities. He used the remainder of the tape to seal what remained of his modesty and slinked off, not to be seen again for the rest of the afternoon.

There was no need for all of the triple protection. Norm and I had sealed the hives, and there were few free-flying bees. All these guys needed to do was carry the hives from the truck about fifteen feet to the squat wall along the dock where they were placing them until they decided where on their roof they were going to keep them. All was well until halfway through the transfer when one of the fellows, for reasons known only to him and G-d, started unscreening the hives that had been placed on the wall next to the water. This sent the bees, all of whom had just been rattled and bumped for a couple of hours driving in stop-and-go traffic on Interstate 95 and bouncing down the Brooklyn-Queens Expressway, furiously flying out of their homes. Chase was again ensconced in his Stay-Puft Marshmallow Man–like suit, but there were several people— curious observers invited by Chase, I suppose—who hovered nearby wearing no protective gear at all. The second half of the transfer was tough to watch as arms flailed and witnesses scattered. Spring had by now indeed sprung and the bees were active and making their presence known; the season was very much upon us.

The remainder of the hives were offloaded soon enough, and Ben and I shook gloved hands, exchanged knowing looks through our veils, laughed, and parted.

Ancient Egyptians considered it good luck to meet a swarm of bees on the road—though a non-beekeeper might wonder what bad luck looked like if a swarm of bees was good luck. For myself, I never put much stock in luck. It may not be an original thought, but the harder I work, the luckier I get. In dealing with honey bees, one needs to be respectful, cautious, attentive, and diligent. "Hardworking" does not even begin to describe the ethic needed to manage colonies of honey bees on a scale beyond that of the hobbyist. So with spring underway, dandelions splattered all over the lawns, the days getting longer, and the bees flittering hither and yon, if one chooses to work with honey bees, one needs to remember the work and perseverance this choice demands. Shortcuts and sloppy labor will result in painful problems like stings. The bees will immediately let a person know their displeasure, a quality that I admire. I have absorbed my share of stings, enough to keep me humble and to remind me, when I get a bit puffed up, that I still have plenty left to learn from these celestial creatures. More than dispensers of stings, honey bees are communicators of love. Sometimes love stings.

APRIL

I'm a failure as a woman. My men expect so much of me because of the image they've made of me and that I've made of myself, as a sex symbol. . . . They expect bells to ring and whistles to whistle, but my anatomy is the same as any other woman's and I can't live up to it.

—MARILYN MONROE

When Lucifer, the most beautiful angel, was cast out of heaven, he took a third of the angels with him upon his departure. When bees swarm, about a third to half of them alight from the hive to seek out a new kingdom in a manner similar to that of the fallen archangel (but with more sweet than evil intentions).

In our neck of the woods, southern New England and New York City, April can be the start of swarm season, though it's certainly not the height of it, as most swarming usually takes place between Memorial Day and Independence Day. A swarm is the way that colonies propagate themselves. Or, more simply put, it's when one colony of honey bees splits into two. There are a couple of reasons why this may happen. Perhaps the hive was strong enough to survive the winter but is not well tended in the spring—meaning not given the room it requires to expand. As the egg laying of the queen increases and the colony repopulates quickly, without proper care it may easily and quickly become overcrowded, and swarm by April. Even without overcrowding, if the colony is poorly ventilated, it may swarm as well.[*]

Evolutionarily speaking, swarming is a sign of a healthy hive, but it is not necessarily ideal for half of the bees to leave the hive on their own terms: A good beekeeper will instigate an artificial swarm—that is, split the colony in two, preventing the loss of bees and increasing his or her apiary.

Prior to a swarm, the queen will do what a lot of women do before a big event like a wedding or a class reunion—she'll fast, and at best she can lose as much as 30 percent of her body weight, because the queen will need to be light to alight from the hive. The queen has mandated help for this—it is really her daughters' idea as they just feed her less to help her achieve their goal. Whereas on some days she will lay thousands of eggs, at this juncture the queen will cease laying eggs for a spell. When she eventually leaves her home, in addition to abandoning about half the population of younger bees, she'll also leave several populated queen cells. From these

---

[*] For detailed information about swarms, refer to Thomas D. Seeley, *Honeybee Democracy* (Princeton: Princeton University Press, 2010).

queen cells will emerge new virgin queens. Hopefully one of these virgin queens will be born successfully, have a fruitful mating flight, and become the new mother of the colony. At this time the oldest foragers are already searching for a new home outside of their current one.

The transition will not be bloodless (though honey bees don't actually have blood; they have hemolymph). There will be a dozen or so fertilized eggs that have been groomed to be queens, and whichever emerges from her cell first—sixteen days after being placed—shall wear the crown. Her majesty's first order of royal business is to visit the still-incubating would-be queen bees and sting them to death as they lie helplessly cloistered in their cells. The queen's cell is elongated, like a peanut, not the small, flat cell of a worker bee or the bulletlike protrusion of a drone. The first queen to emerge from her cell will either sting her competitors to death through what will shortly become their peanut-shaped coffins, or bite a hole in the side of each cell so that worker bees will finish the job and dispatch the would-be royals. On the occasion that two or more queens are born more or less at the same time, they will detect each other through scent, locate one another, and battle to the death. Sometimes a virgin queen is born and remains in the hive along with her queen mother for a short time. The worker bees keep the two queens, the old established queen and the newly born one, apart from one another. Or not! Most people believe that a colony cannot tolerate more than one queen at a time, but this is not the case. Many colonies do just this. But in the circumstances being described here, this is a temporary situation leading up to a swarm.

So if and when they deem it necessary, the workers make

a collective decision to depart from the hive. As a final preparation, they load themselves up with honey for the journey.
They usually do not go too far at the onset. Generally, if one
is available, the bees will fly to a tree no more than twenty feet
from the original hive. There the bees will cluster, in a more
loose-knit manner than in the winter cluster, around the queen
bee. This cluster usually ranges in size from a grapefruit to a
soccer ball. Since the swarm has no brood to protect and they
are full of nourishment, they are at their most docile and least
likely to sting. While the majority are gathered there on the
limb or trunk of a tree—or, if in Manhattan, perhaps on the
umbrella of a hot dog vendor's cart in Times Square, the side-
view mirror of a car parked along the West Side Highway, the
front door of a high-end restaurant near the downtown federal courthouses, a fence on the deck of the USS *Intrepid*, a
traffic signal, a fire hydrant, or the center field wall at Yankee
Stadium just hours before a big game, to name but a few
places where city bees have swarmed—the scout bees head
out in search of more suitable living accommodations.

Scout bees are worker bees with the most experience foraging. As many as one hundred of them take on this role.
These experienced pilots strike out from the cluster and seek
out an easily defensible, dry, cool, dark cavity somewhere—
perhaps a hollowed-out tree, maybe a void in a section of a
stone wall—where the group could set up a new home. Upon
finding a suitable new abode, a scout will return to the group
and indicate the coordinates of the option via dance. Other
scouts will also return with options, also relayed via dance.
Scouts will confer with other scouts, maybe collectively check
out one another's findings, and eventually, be it within a few

hours or a few days, like Quakers, they will come to a collective agreement. They will fly as far as a mile away to their new home and commence building honeycomb and carrying out the usual tasks associated with their lives. If a beekeeper is able to be in the right place at the right time, he or she may capture the swarm as they are in that limbo between their former home and the new home they seek. Those bees can be relocated into a new beehive and put to work for the lucky soul who captures them.

The life of the queen may not be all it is cracked up to be. A productive queen must lay up to two thousand individual eggs per day during the peak season of early spring. This effectively boils down to laying her own body weight in eggs every day. She determines the gender of each one by fertilizing or not fertilizing it as she plops it down into the cell, by inspecting each cell before she lays into it. That is, if the workers construct her drone cells, she lays drones, and if a worker cell, she lays workers. Again, the workers are in control even of the gender of their future siblings. The queen is responsible for the welfare of the hive in continuing the population of the colony. Though she is in charge of the hive in one sense, the workers will rise up, and destroy and replace her should she falter—that is, if she fails to produce the required number of viable eggs and queen pheromone for the community. So she is a queen, but in a gilded cage. She is more a slave than a monarch. In the beehive, the workers are truly in charge of the means of production and have all of the power. Marx would have approved.

Usually, a queen leaves the hive only once in her life, for her mating flight. After she has her one afternoon of multiple trysts in the sky, she is relegated to a lifetime of reflection and toil; years of pushing out babies to ensure the survival of the hive. She is, in a sense, a prisoner of her own daughters. Unless she swarms, she will not see daylight until the beekeeper tips back the lid of the hive and pokes around during an inspection. Fortunately, the queen is modest and would just as soon not see bright light, which irritates her and sends her scrambling for a dark corner quicker than Blanche DuBois.

When the hive needs a new queen, it creates several of them from ordinary eggs. For though a queen is born a queen, when she is implanted into a cell as an egg, she's no different from any of her sisters. When a new queen is required, a few wee larvae less than thirty-six hours old are selected by the workers and cultivated expressly to wear the crown. If the need is not the result of a swarm or an impending swarm, it's likely to be due to the current queen's poor egg production, the beekeeper perhaps accidentally squashing the queen bee, a blue jay gobbling her up as she returned from her mating flight, or some similar calamity. In this case, the replacement process is called supersedure.

Under normal hive conditions, all larvae are fed a diet of royal jelly for at least thirty-six hours. In fact, all eggs are fed one thousand times per day and bathed in royal jelly, completely saturated in the nutrient-rich substance, which is secreted by the hypopharyngeal gland in a young worker bee's head. After that enriched soaking, for most eggs, it is a steady diet of honey and pollen for the remaining days until emergence from the cell. Unfertilized larvae become drones (males), spending twenty-four days in a cell, while more become workers (females), but without functioning ovaries, spending twenty-one days in the cell. But queen bees, which start off as larvae selected by nurse bees, are fed only royal jelly, and as a result, they grow bigger, more quickly, and mature in far better condition than the sterile worker bees. Plus, their diet of royal jelly awards them functioning ovaries.

Based on the anatomy of the worker bee, one may wonder how a queen is able to sting and not die. Unlike the stinger of the typical worker bee, or the phallus of the drone, the queen's stinger is not adorned with barbs, and she may sting as often as she likes. Yet she will sting only another queen. This allows beekeepers to handle the queen without fear of being stung, and unsatisfied worker bees to usurp power and commit regicide without fear of immediate royal retaliation.

Thanks to her steady diet of royal jelly, the queen can live for two to four years; workers and drones live for six and eight weeks respectively, on average. However, the queen generally starts to slow down in her egg laying after the second year, so much so that if a beekeeper does not replace her, the bees may take it upon themselves to create a batch of new queens. Some beekeepers are game to allow nature to take its course and let their colony requeen itself. I am of the school that believes the hobbyist beekeeper, particularly at my latitude, is better off buying a properly mated queen from a reputable breeder. But consensus is not common in the beekeeping community. An old and true adage is that if one were to ask ten beekeepers for an opinion on any beekeeping matter, there would be eleven different responses—and each would be delivered with certainty and exactitude. Like in all specialties, there are many opinions held by different beekeepers.

We must remember that we have taken a mild departure

from nature when we decided to put honey bees in boxes and bother them all year long, prodding at their homes with our metal hive tools and filling their faces with exhaust from our smokers. So while many market themselves as "natural" bee-keepers, it makes no sense to imagine that one method is much more "natural" than another, if we are talking about humans trying to keep bees in boxes, whatever the configuration. So, to have the bees engender their own queen rather than to purchase one of battle-tested genetics might not be ideal for the hobbyist hoping to harvest. Aside from all else, allowing the colony to mate a queen on its own would mean a long delay in production, with waiting for the queen to hatch, go on her flight, be impregnated (perhaps with drones of questionable genetics), return, lay, and then for those eggs to develop and for those workers to mature enough to be able to leave the hive and be useful on the outside. This would not only adversely affect the excess honey to be harvested, but could negatively impact the hive to the point that the delay could render it impossible for the colony to gather what it needs to survive the winter. Say one lets the bees do things their way. The bee-keeper might end up with two queens who battle it out to the death. The better fighter may win that regal battle, but be a disaster when it comes to laying. This always vividly reminds me of my first marriage.

Each year in early April, often over Passover and/or Easter, my father, brother, and I drive one thousand miles in each direction to pick up millions of bees from Wilbanks Apiaries, a large commercial apiary ("commercial" meaning more than one thousand colonies) in Claxton, Georgia. Lately we have also been sourcing hundreds of packages from an apiary in Northern California called Olivarez Honey Bees, which specializes in, among other types, Carniolan honey bees.

These bees are not all for us. Thankfully, we have never needed to replace all of our hives. Currently the national annual average for bee loss is above 30 percent attrition, and we are usually near that mark. But we continually get new clients, or we may want to bolster weak hives, and so each spring we replenish many of our colonies. Mostly, though, we acquire these packages to sell to other beekeepers in the five boroughs whose own hives did not overwinter successfully, or who are new to the beekeeping game. We also procure them for beekeepers all over New England, both hobbyists (who have a few colonies mostly for their own pleasure) and sideliners (who usually have a day job but have a greater number of hives than hobbyists and who try to make a side living via their honey bees), who meet us at a New York or Connecticut location and pick them up from us there.

Most beekeepers who are more than hobbyists need to make their living not only through the obvious sale of honey but through sales of bees and beekeeping equipment, the provision of pollination services, by giving talks or classes, or through the removal of colonies of bees that have taken up residence in a structure. In other words, through whatever manner of sales and services that might help them eke out a living. It is often said that a boat is a hole in the water surrounded by wood into which you pour money. Beekeepers say that the way to make a small fortune in beekeeping is to start with a large fortune and spend it as your hobby increases. That sounds about right to me.

Like honey, honey bees are sold by weight. Most packages are three pounds of bees, which at four thousand or so per pound means about twelve thousand bees per three-pound package. Also in the container is a small queen's cage, in which a single queen resides, protected over the long journey from the workers who may not immediately recognize her as their elected official. Her cage usually houses a few workers as well, who act as ladies-in-waiting, tending to her needs. Rather than instantly injecting her into a colony that might perceive her as hostile and kill her, the beekeeper often plugs the cage with candy, like confectionary sugar or marshmallow, so that when it is placed in with the general bee population, the queen will be released slowly as the workers eat away the sweets. This delay allows time for her pheromone to spread, helping the bees to accept her as queen. The package itself is traditionally a wooden box, just slightly larger than a shoebox, with screened sides that comes replete with an inverted can of sugar water to allow the girls some nourishment during their journey. In the past few years, some outfits have begun using plastic box cages instead of the traditional wood and screen. We don't prefer them for several reasons—not the least of which is we don't need more plastic floating in the oceans. But I find that bees respond better to natural materials.

Not all beekeepers wait for someone to appropriate the bees for them. Some people have them shipped via the post office. Yes, packages of honey bees have been shipped via the United States Postal Service for more than a century, and still are to this day. Most breeders don't recommend this since temperature control is not optimum and, generally speaking, postal workers would rather have nothing to do with bees, which may lead to less-than-ideal care. But it is done. One example comes from the 1970 Sears, Roebuck catalog, which, at least according to *The Christian Science Monitor*, listed packages of honey bees for "\$7–\$12 each, depending on the strain." It elaborates: "In fact, Sears ships up to three tons of bees a year to mail order customers." That's a lot of bees, considering they are usually sold in packages of three pounds.

It should be no great surprise that packages of twelve thousand honey bees each can be shipped via USPS, given that when the USPS introduced the parcel post service in 1913, it was used on occasion to ship babies and children. There are at least two recorded instances. One, involving a boy shipped from Stratford, Oklahoma, to Wellington, Kansas, was reported in *The New York Times* about a year later:

> Mrs. E. H. Staley of this city received her two-year-old nephew by parcel post to-day from his grandmother in Stratford, Okla., where he had been left for a visit three weeks ago. The boy wore a tag about his neck showing it had cost 18 cents to send him through the mails. He was transported 25 miles by rural route before reaching the railroad. He rode with the mail clerks, shared his lunch with them and arrived here in good condition.

As a parent, I can attest to the fact that children are much more difficult and challenging to manage than honey bees. Compared to a screaming baby with a bursting diaper, a box of thousands of venomous stinging insects sounds positively tame to me. In any case, the practice of shipping children via the mail ended by 1920, as *The Washington Herald* exclaimed in its headline: CAN'T MAIL KIDDIES—DANGEROUS ANIMALS. The post office, in its wisdom, had finally ruled that children were not "harmless animals," and because of their potentiality for danger may not be mailed as parcel post. "By no stretch of imagination or language," said the ruling, "can children be classified as harmless, live animals that do not require food or water." But bees were and still are accepted for shipment.

So in any given year, the anticipated arrival date of a package of honey bees in New York from Georgia might be, for instance, the first Friday in April. People place their orders, plans are made, anticipation builds. But we are dealing with nature. If one plans an outdoor wedding for June 1 and it rains all day, plans for the altar may need to be altered, and everyone understands that provisions must be made for this. We gently hammer it into the heads of all beekeepers and beekeepers-to-be that weather is a critical factor in the timing of the actual arrival of the bees. In the end, the bees may not be ready when one wishes them to be ready. Managing people's expectations is routinely more difficult than handling the bees themselves. Particularly when it comes to our valued but often mollycoddled customers in New York City, the notion of letting something like weather interfere with their plans is often too much for their delicate sensibilities to fathom, and they often exhaust themselves with a long series of verbal gymnastics and "what if" hypothetical arguments designed to bring them their desired results. To no avail.

We do not control nature; we merely cajole the circumstances around it, and usually not very well. One could prune a tree. Mayhap bend its limbs to cause it to grow a certain way. In the end, that tree still needs to have a root system in the ground and it still needs water and nutrient-rich soil and it still needs sunshine. It needs space to grow. There's only so far we can force nature—in this case, bees—and have things work for us the way we would prefer them to. So while we are able to control and incentivize honey bees to do certain things—like, thanks to their uncanny olfactory abilities, detect bombs, or even, by virtue of a patient's breath, detect tuberculosis and certain types of cancer—there's only so much we can manipulate nature and not destroy what it is we are trying to cultivate.

When mid-March torrential rains disrupt queen breeding, package production will slow down. Similarly, if the weather is too dry or too cold, bee populations will be insufficient to create the packages requested, and the arrival date will be pushed back one or two or three weeks or even more. Of course, we advertise this fact just as strongly as I tried to impress upon the Ugandan farmers how important it is not to eat banana mash in the morning before visiting an apiary. With nothing like the Ugandans' excuse of a tricky language barrier, my fellow New Yorkers often fail to follow the plot.

My cohorts of New York–area beekeepers are generally cognizant and understanding of these truths. Some, however, are not swayed by logic or facts when put up against their own desires for things to be as they wish them to be. Herein the humble honey bees have many lessons to teach us. Perhaps aside from humility, which is the first and most important lesson honey bees drill into their keepers, one of the best lessons they have for us is patience. Honey bees can teach us to be patient through a Gandhian style method of civil disobedience. Meaning, the bees are going to do what they want to do, and we need to adapt our ways to their choices if as a joint endeavor we are to be successful.

One of the many things I love about honey bees is that they will always let a person know where he or she stands. A brief moment here for a personal example using yours truly and my beloved wife, Yuliana. As difficult as those who know me might find it to believe, perhaps, on any given day Yuliana might seem a bit miffed with me. When I ask her about it, she may look directly at me, smile her lovely smile, and assure me that nothing is wrong. If I have my doubts, persist in gently asking, and the response remains the same, I might be lulled into believing that, in fact, nothing *is* wrong. But an hour or a month later, I may well find out that Yuliana was actually upset about something of which I, dumb drone that I am, had been blissfully unaware. In these cases, the pain is a long, slow, drawn-out burn.

The point is, the bees will not put one through all of this anguish the way a human might. With the worker bees there is reliable and unqualified clarity. Judgment and retribution are quick. If bees are upset with someone for whatever reason, they will make that displeasure known posthaste. They may hover around one's face, they may mob a veil and attempt to get in, or they may simply sting. Unlike with some human females, there is no ambiguity. Rudyard Kipling was spot-on when he wrote "The female of the species is more deadly than the male." There is something wonderfully absolute and pure in knowing that the bees will always be direct. Nature has much to teach us.

In terms of distributing packages of bees in April, these days we give people two options. One is to pick them up at my parents' home in Norwalk, where my mother, Polly; my niece, Megan; and my nephew, Patrick, check names against a master list and distribute the bees until they are all gone. My mother comes out to greet us when we arrive at her place, usually around two A.M., and we offload as many packages as she needs for her pickups. We set up a tent for the rain or the sun and make sure she is all settled with her lists and highlighters. We notify customers from the road that we are on schedule, and they start arriving at dawn to collect their girls.

But before we reach as far north as Connecticut, some of us meet up at a highway rest stop or perhaps near Yankee Stadium where we transfer around 250 packages into a smaller truck and drive into Manhattan, while the other continues north to New England. Once in Manhattan, in collaboration with the crew of the New York City Beekeepers Association, we make our early morning deadline, usually six or seven A.M., to meet the city-dwelling beekeepers who are anxiously awaiting their own packages of twelve thousand fresh new bees. In both locations, city and country, the exchange is generally a happy occasion. People are pleased to meet their new bees and to see familiar human faces among the crowds. There are plenty of reunions, as we have many repeat customers from the five boroughs and the neighboring states, and people value the opportunity to talk about their yields, display samples of their harvests, and enjoy some good bee chat with like-minded folk.

Sometimes, and this is particularly a thing in New York City, where people complain as if it is a competitive sport with big cash prizes, people whine that there do not appear to be a full twelve thousand bees in their packages. We invite them to count the legs and divide by six. No one has yet taken us up on that. But we have had people get tremendously upset when there is a delay due to rain, cold, or some other issue beyond the control of mere mortals. As explained, poor weather means that the queens cannot be mated, and so the populations cannot grow to meet the need to create the packages in a timely fashion. At these times the bees may end up being ready on, say, the weekend of Easter. When folks "get ugly," as my Alabama-born mother would say, about all we can do is hope they remember their Scripture (bless their hearts):

"Therefore be patient, brethren. . . . The farmer waits for the precious produce of the soil, being patient about it, until it gets the early and late rains. You too be patient" (James 5:7–8). Or perhaps, "And He said to them, 'Which one of you, will have a son or an ox fall into a well, and will not immediately pull him out on a Sabbath day?'" (Luke 14:5). But I cannot win—those who rally against me that their bees disrupted their Easter plans do not want Scripture quoted to them. They want their egg hunts, ham, and family gatherings, and perhaps, a visit to church in a bright showy hat and not a beekeeper's veil.

As beekeepers we must understand that even on inopportune days, sometimes we need to do things that need to be done. Like changing a flat tire, or retrieving our bees, or dredging a fallen child out of a well on the Sabbath. It can be considered a lesson in beekeeping that we must bend to the schedule of the bees and to the elements. Unfortunately, the bees are ready as soon as they are ready—meaning, alas, we cannot postpone a week, or even forty-eight hours. When the season is upon us, it begins.

In Connecticut, the handoff from my mother, Polly, is

rather laid-back as there is an open window over two days when people may pick up their packages during daylight hours.

In New York City, the package pickup schedule is more compressed, as we cannot sit on the side of the road with millions of bees for too long. When the bees arrive, clusters of "hobo" bees hang on to the outside of the packages, creating confusion and sometimes horror in onlookers who believe that the bees are escaping. These hangers-on have been gripping the screen with their six feet for the last thousand miles, smelling a queen within and wishing to join the party. But to the uninitiated, the hobos appear to be escaping bees, and may disturb those walking to work. Also, some years the heat may stifle the bees. Most important, we have to tend to our own colonies. So we generally have a set time and place, and people form a line and get their bees, exchange a few pleasantries, then head for their bicycle/subway/bus/car to get themselves and their tiny charges home.

As urban beekeepers, we have certainly come a long way since the days of hiding our beehives on rooftops, painted to resemble chimneys or in shades of gray so they blend in with rooftop machinery. During what I jestingly refer to as "the Bitter Years" of illegal beekeeping, the New York City distributions used to take place in clandestine locations: a community garden in Prospect Heights on the corner of Vanderbilt and St. Marks, or a quiet spot near a synagogue in Forest Hills, Queens, for example. For several years, illicit distribution was a fun part of the Union Square Greenmarket, until we outgrew that space and started to disrupt the farmers' market with people seeking their bee packages, and wayward hobos of both the insect and human form.

The packages were disseminated publicly in Columbus Circle for a few years once distribution was no longer a clandestine operation. More recently, we have made use of Bryant Park, behind the main branch of the New York Public Library, to distribute to the hundreds of people who turn up to stand in a long line to receive their packages of bees. In 2018, distribution there garnered a *New York Times* front-page story focused on the congeniality that urban beekeeping and beeks contribute to city life.

Now and again during distribution, some bystander will become disturbed and call 911 for no good reason. The NYPD is required to respond. But any potential police interference halts as soon as the bees rear their heads; the cops don't have enough cuffs for the six-legged creatures, and it is amazing how one of the largest police forces in the world can seem so powerless in the face of a few flying insects.

Jesting aside, the NYPD and its Emergency Services Unit have always worked with us in a friendly and positive way, calling upon us to deal with these insects on a regular basis. Not that the NYPD doesn't have staff fit and able to handle bee disturbances. In the late 1990s, up until he effectively talked his way out of a job, Anthony Planakis—known as "Tony Bees"—was the fellow they used for beekeeping issues. Following Tony's departure from the force, a detective from counterterrorism named Dan Higgins took over the semi-official role as department beekeeper. He held it until he transferred out of the NYPD to take a position closer to home in Westchester. That's when Officer Darren Mays, who lives in rural upstate New York but whose precinct is in Queens, took over the job. He now shares it with Officer Michael Lauriano of Long

Island. So, as of this writing, there are two NYPD officers who help this city of millions deal with the swarms and honey bee situations that crop up now and again. I have worked with both of them, and have been friends with all.

There is always something interesting—at least to beekeepers—happening with honey bees in New York City. One spring day some years back I was asked by a producer to appear on a television show called *Cake Boss,* a reality show dedicated to cake making. In the United States, the majority of the population seems to not only enjoy eating foods that are bad for them in great abundance, but watching others do so, too, so this show was wildly successful. The *Cake Boss* people wanted to film their star, Buddy Valastro, of the family-owned business Carlo's Bakery in Hoboken, New Jersey, following me onto a rooftop in Manhattan to obtain honey from some hives there. The way we portrayed it all—harvesting the honey one bottle at a time—is not the way honey harvesting is done, but there are a lot of liberties taken with making these sorts of shows, as I quickly came to learn. In the aired episode, the crew took the honey to their shop in New Jersey to become an ingredient in a cake. Finally, in what was meant to be another day, there was a reveal at a community garden and a group of us were gob-smacked by the magnificent cake and in shock and awe of the tremendous talents the bakers displayed in creating it. At least that is how the program was edited and how we were instructed to react. In reality, the finished cake was actually a bit odd. It was also a prop and, as such, inedible.

I told a beekeeping acquaintance of mine, an aspiring

actor named Mickey, about the upcoming *Cake Boss* shoot, and he begged me to include him. Between rare auditions, Mickey had been introduced to beekeeping in 2005 in the Clinton Community Garden on Forty-eighth Street between Ninth and Tenth avenues, coincidentally where I currently take care of the beehives.

"I [had] just started taking dance classes, and I heard honey bees love to dance, and they do have a little dance that the scout bees have when they go back to the hive," Mickey said in an interview on National Public Radio that year. It is true that honey bees use dance to communicate. In fact, there are two types of dances—the waggle dance and the round dance. The first is used to communicate to other bees the distance and directions to sources of nectar, water, or other colony needs. The second is more or less used for the same thing, but for locations much closer to the hive.

Once his interest was kindled, he worked alongside and learned about beekeeping from a man named Sidney Glaser. Glaser had learned beekeeping while working in Paraguay with the Peace Corps. There, among the Guarani, Glaser developed a love of and appreciation for the honey bees, even though most of the bees there were Africanized and therefore much more defensive and challenging to handle than those in North America. A decade later, working with the honey bees in what was still a dicey Hell's Kitchen must have been a cakewalk compared to the belligerency of the honey bees in South America. So Mickey learned from someone who had truly made his bones in the beeyard.

Mickey introduced himself to me sometime thereafter. We saw each other at various bee functions, became pals, caught swarms together, and did a cut-out or two. (A cut-out is when honey bees have entered a space within a house or other dwelling, built comb, and fully taken up residence. The colony needs to be removed by opening up the structure and removing all of the comb and the bees. It is messy work.)

At the insistence of Mickey and his wife, I visited their apartment, where she gushed about how much she loved *Cake Boss*, begged me to get Mickey in the shoot, and started to tell me all about her pole-dancing class. Hoping not to hear more about that last item, I said, sure, I would ask if Mickey could attend the filming with me, and if the producers agreed, he could assist me with the beehive inspection and make his debut on the small screen. I approached the producers with the idea, and though they were initially against any changes to the plan, I knew it was important to Mickey. So I gently pushed, and they relented.

Prior to the legalization of beekeeping in New York City, a reporter from ABC's *Nightline* visited me at a rooftop apiary belonging to my friends Peter and Deborah Dowling, where they filmed me for my own television debut. I was excited. The night that the segment was to air, I went to a bar where there were a couple of televisions. I asked the bartender to tune to the appropriate channel, and he did so with about as much interest as I had in helping him wash lipstick-smeared glasses. In my mind, when my big moment came on, it would be like in the movies. The bar would be crowded, I would sit there acting disinterested, nursing my drink. Someone would look at the screen, then me, then the screen and back at me again. Maybe nudge a companion and nonchalantly bend their head in my direction, then dart their eyes to the television screen. Then someone would say, "Hey! That's *you!*" And I would act as if I did not care, like it was an everyday occurrence. Or something like that.

Needless to say, though I enjoyed the spot, no one noticed me, and luckily I had just enough dignity not to start pointing it out to people. I paid for my drink and left, not feeling too disappointed, but a twinge silly.

But before that, after seemingly never-ending back-and-forth changes with the *Cake Boss* crew, we worked out all of the arrangements and finally came up with a date and time to do the rooftop shoot. I showed up at the now defunct Bridge Cafe in the Financial District of lower Manhattan on the corner of Water and Dover streets. Well known as a former haven for pirates, the Bridge Cafe, according to owner Adam Weprin, was the oldest continuously open pub in New York City, as well as being New York City's oldest commercial wood-frame building, dating from 1794. It was also reported to be haunted by the ghosts of prostitutes who had worked upstairs. These ghosts of ill repute supposedly trailed lavender scents behind them as they made their otherworldly rounds in the rickety old haunt. I never encountered them.

On the day we were to film the scene with the Cake Boss at the restaurant, I met Mickey, who was there ahead of schedule and smiling fit to burst, and together we ascended the dark, creaky centuries-old staircase. It was then that we discovered the place hadn't been renovated in who knows how long. The second and third floors were completely without electricity and nowhere near up to code. The owner used the space as a huge warehouse for old paperwork, and also apparently as a safe house for wayward pigeons, as some of the windows were permanently open and pigeon excrement heavily decorated boxes that bore handwritten masking-taped labels that went back four decades.

Though he showed up on time, Buddy the Baker spent about ninety minutes talking on his phone before we could start the filming. When no one seemed to be able or even willing to try to tell him to get off, I started to envision how the rest of the day might go. Still, nothing could have prepared me for what I was to encounter. Buddy was the least of my concerns when I turned to see Mickey, standing just behind one of the beehives, as naked as the day he was born. Nude beekeeping would "set me apart," he later claimed, and "give me more exposure." No doubt both things were true; there was indeed a great deal of pale exposure. I feared for the white balance on the cameras.

While I stood gaping, utterly speechless, Mickey took the outer cover off the far hive and stood on it to make himself taller for the camera. "A little trick I learned," he said proudly, not at all addressing the elephant—or inchworm—on the roof. Mortified, I braced for a shutdown of production.

Somehow the filming went on, with three fully dressed men and one fellow unencumbered by the trappings of fashion, checking the inside of the beehive. To me at least, no one on the production team said a word about the naked beekeeper, nor did Buddy. After a number of takes, we wrapped that scene. The "talent," as the star of the show is called, immediately left, the equipment was packed, and the crew of about eight descended to the cobblestone street below. The crew, Mickey—mercifully dressed once again—and I then headed for a community garden where I was also keeping beehives,

Green Oasis, on Eighth Street between avenues C and D. This was in what was once referred to as Alphabet City and is now mostly known as the East Village. When the show was aired, the viewing audience would be told that the scene in Green Oasis was taking place a few days later, during a gathering of New York City beekeepers who were waiting for the cake that Buddy and his crew were to deliver to the garden. The occasion was intended to celebrate the one-year anniversary of the legalization of honey bee keeping in New York City. There were several dozen people milling about, happy and genuinely excited to be on the show and to see what the ingenious bakers had fabricated.

The inedible prop cake appeared, and it looked like something out of a Winnie-the-Pooh nightmare. Thanks to editing magic, after that "cake" was cut on camera, we were served a different cake, a sheet cake that tasted more or less like every other sheet cake in the world.

A few months later, when the show was set to air, Mickey was a ball of nervous excitement. To witness his television debut, he invited a group of people to a bar in Brooklyn where he had arranged for the television to be tuned to TLC to watch his glory on a huge screen. I wasn't there (I may have been too traumatized from my own experiences at a bar seeing myself on television), so I don't know exactly how many people showed up, though he told me he'd invited everyone he knew and then some. I had a few mutual friends there, so I did not have to imagine the scene—the crowd waiting in anticipation, Mickey breathless and squirming, his wife squeezing his arm in excitement. Drinks flowing. When the program came on, cheering from the crowd began. Mickey smiled and stared at the screen so as not to miss an inch (the inch?) of his debut. The show ran for its full twenty-two minutes plus commercials with every bit of Mickey edited out except for one half second during which he could be spotted stuffing his wide-open mouth with replacement sheet cake in the background at the community garden.

I don't see much of Mickey anymore. Certainly not as much as I saw that day. I learned that sometime later, Mickey had been handcuffed and taken to the Seventy-ninth Precinct station house in Bedford Stuyvesant, Brooklyn, where he'd been tossed into a cell. He ended up being issued two summonses—one for disorderly conduct, and one for violating article 161.03a of the New York City health code, which declares that "a person who owns, possesses or controls a dog, cat or other animal shall not permit the animal to commit a nuisance on a sidewalk of any public place." In this case, it was "other animal." Namely, bees.

Tony Bees, the go-to cop for beekeeping issues at the time, had been called to Madison Street in Brooklyn to deal with tens of thousands of bees that had escaped from a parked car, which turned out to be Mickey's. Ironically, Mickey was teaching a beekeeping safety course at the time of his arrest. On a sweltering hot day, he had chosen to leave boxes of poorly contained colonies of bees to roast in the stifling heat inside his car, with the windows barely cracked. Some of the plastic packaging melted just enough to make escape possible, and thousands of the bees hovered around the spot, creating an unsafe condition in the area surrounding the car and along the sidewalk.

Joining half a dozen uniformed police officers summoned to the scene, Tony called me once he was there, as Mickey had dropped my name in an attempt to mitigate his circumstances.

"Andrew and I are good friends," Mickey told Tony, smiling his most winning smile. Tony called and asked me if that were the case.

"I know him. I'm not quite sure I would want to post bail for him or anything like that," I responded honestly, not knowing the details of the situation.

"What should I do with him?" Tony asked.

"What do you mean?"

"You want I should arrest him?" Tony briefly explained the carelessness with the bees, the distress of the neighbors, the whole ugly scene. It was bad for the bees, and worse for New York City beekeeping's recently hard-won legalized position. I could not defend what Mickey was doing.

"I wouldn't hold it against you. It's your call."

"Got it."

Phone clicks, followed shortly thereafter by handcuffs clicking.

Mickey's claim to fame had gone from an instant of chowing down mediocre cake on television to being the first beekeeper arrested for beekeeping-related activity in the history of New York City. In the end, he was proud to see his name in *The New York Times* when it reported the fiasco, and he shared the story far and wide. They say there is no bad publicity, and for him, it seemed to be true. Unrelated to all that, he has taken his wife and three daughters and moved to New Jersey, so we don't see too much of him anymore. He still keeps bees, from what I gather, and is still trying to perfect his bee dance.

MAY

There is no doubt that a certain similarity can be seen between bees, which sculpt the honeycombs out of a soft substance (wax), and sculptors, who do as much with the same material (or equivalents). . . . Is it therefore so surprising that, in this context, the artist should be associated with the self-sacrificial bee?

—JUAN ANTONIO RAMÍREZ,
The Beehive Metaphor: From Gaudí to Le Corbusier

Hallie Tew, born Hallie Gustavson, was a dog walker, cat sitter, and bee enthusiast based in Greenpoint, Brooklyn. She grew up in Wisconsin, where she met her future husband, Alex,* a painter (canvas, not walls), and married him, grasping a head of broccoli in lieu of a bouquet of

* Both of these names have been changed.

flowers as she walked down the aisle at their wedding. Hallie and Alex moved to Brooklyn around the start of the new millenium, and began their joint lives. Then she worked in an office in Manhattan and was very stylish, often dressing in her grandmother's vintage outfits. From her bathroom window she watched the Twin Towers burn and fall. Alex, tall, bald, and bearded, is a modish dresser whose studio is also located in Greenpoint. In the early 2000s they were, by all accounts and appearances, a typical young couple from America's Dairyland living the big-city dream along with their cats in their narrow ground-floor Brooklyn apartment.

From a young age Hallie had been interested in honey bees, even dressing up as one for Halloween one year in elementary school. One fateful day, as an adult living in New York City, much to her dismay and horror, Hallie learned that she was allergic to honey bees. Not to cats, dust, mold, peanuts, tree pollen, ragweed, or anything else—only honey bees. "Bees make me break out in hives!" she lamented. Her corny joke aside, Hallie was heartbroken. Still, she would not be disabused of her longing to keep bees. She dove headfirst into urban beekeeping despite the dangers.

Backing up a bit, Hallie and I met at a beekeeping Meetup at a diner called Odessa, on Avenue A right across the street from Tompkins Square Park, where we joined about two dozen others with an interest in the sweet insects. This was during the Bitter Years, when beekeeping was still illegal within city limits. (Not that there was ever a posse sent from the city to crack down on clandestine beekeeping.)

The get-together was arranged via a now-defunct online group called Brooklyn Beekeepers Meetup that sought out like-minded individuals who were passionate—or at least curious—about bees. Hallie and I found ourselves seated across from each other. When I left the diner to return to my farmers' market across the street, she made a point to go there, speak to and befriend me. As a result of a friendship that then went a bit too far, and her husband's disapproval of another man pollinating his wife, Hallie became the subject of what is arguably her husband's greatest and most personal work of art: a seven-foot-tall oil on canvas that features Hallie painted lifesize, full height, lying on her back, eyes closed, dead, naked as the day she was born, and covered in and surrounded by honey bees, honey, and honeycomb patterns—entitled *Death Portrait of Hallie Tew.*

This isn't the only time that honey bees and art have collided, though perhaps it is one of the more morbid ones. And despite her death portrait living on display, Hallie, thankfully, is alive and well and still beekeeping. When confronted by her husband, she confessed our sticky liaison to him. He was displeased. He took the small figurine that had once adorned the top of their wedding cake—the two of them as bride and groom, replete with her gripping the head of broccoli—and smashed it. But, she being his muse, he also channeled that emotion into what is arguably his greatest, or at least largest, piece of art, now for sale at a reputable downtown gallery for tens of thousands of dollars.

Here in New York City we aren't immune to the collaboration of the honey bee and the artist and creative types of all stripes. Even the very landscape of New York City is lightly sprinkled with honey bee acknowledgments and tributes. Skeps—old-fashioned beehives woven like baskets in which honey bees traditionally made their homes prior to the so-called modernization of beekeeping in the 1850s, even though wooden boxes had begun to replace skeps as early as the 1820s in Stockbridge Village, Massachusetts—were often used by banks as symbols of stability and safety. There is a huge stone skep hidden away in one of the side pockets in the Children's Garden in the Brooklyn Botanic Garden; before its home in the garden, this particular skep, made in the 1890s, was an architectural detail on the Brooklyn Savings Bank at the Pierrepont Street entrance. When the building was demolished in the early 1960s, the "historic granite emblem" was transferred to the BBG "in a light snowstorm on February 10, 1964," and hasn't budged since. Another skep, this one metal, still adorns a grand old former bank on the northwest corner of Fourteenth Street and Eighth Avenue in Manhattan.

In a more modern example, there is a historic subway station in Sunset Park, Brooklyn, where an artist commissioned by the city in 2012 created seven-foot gates of honeycomb crawling with brass honey bees. Since then, no one seems to have remembered or bothered to polish them (perhaps that is part of the plan). The Church of Jesus Christ of Latter-day Saints, which uses the skep as one of their symbols, has honey bees emblazed in stone on the sidewalk outside of their house of worship on the Upper West Side on Sixty-fifth Street and Columbus, and three-dimensional honey bees carved on the wood paneling of the elevators in that building. Mormons are big fans of the honey bee, and believe that "it is a significant representation of the industry, harmony, order and frugality of the people, and of the sweet results of their toil, union and intelligent cooperation." It's no surprise, then, that Utah, founded by Mormons, adopted the honey bee as the official state insect.

The Broadway-Lafayette subway station in downtown Manhattan has neon honeycomb buzzing on display overhead as one uses the stairs to get to or from the trains. David Bowie would have seen these when taking the D or F train to his multi-level penthouse above the Mulberry Street branch of the New York Public Library, from which he could have easily looked across the street and down into the cemetery and seen my seven beehives—one for each of the seven deadly sins—in the cemetery of St. Patrick's Old Cathedral, which began interring the Catholic faithful in 1785.

This gorgeous old church, the only designated basilica in New York City, was the seat of the Archdiocese of New York until 1879, when that designation was transferred to the "new" and better-known St. Patrick's Cathedral uptown on Fifth Avenue. The grand interior of Old St. Pat's was used for the baptism scene in *The Godfather*, and in real life, the FBI used the church as a lookout point to surveil mob bosses in what was then a mafia-saturated portion of Little Italy.

More interesting to me, though, was the Bowie connection. Like millions of others, I have been enamored of David Bowie's music and persona since before I was a teenager, and the fact that I was so close to his abode on a weekly basis thrilled me. At least once while working those cemetery hives, I sang Bowie's "Please, Mr. Gravedigger" at a not-so-quiet volume in the hopes that it would spontaneously draw Bowie's gaze to my apiary from his magnificent apartment. Alas, and of course, it did not work. But I think I did once hear a window slam shut from somewhere up above.

There are many more examples of bee references among us. Allusions to them are found in art, advertising, religion, folklore, and mythology. For a beekeeper like me who looks for them—or, maybe, cannot avoid seeing them—the visual references seem downright pervasive, from the hexagonally shaped stones that encompass Central Park and Union Square, to graffiti on various buildings in the outer boroughs and largely unnoticed architectural details on sidewalks, buildings, and walls. And somehow, the humans who make up the complex fabric of this city seem to weave bees into my life in ways that I could not imagine on my own.

Being a somewhat well-known, or at least accessible, beekeeper in New York City, I am approached by all kinds of people who have ideas for projects connected to honey bees. Sometimes they want bees; sometimes they want me. Often it's a bit of both. Some are legitimate and worthy of attention, like when a car company wants to show me driving from apiary to apiary, delighted to be behind the wheel of their wonderful Mercedes-Benz Smart Car. (Mercedes-Benz used a white and green electric car for the shoot for the day, "a real missed opportunity not to use a yellow and black one, like a bee!" suggested my good friend Hope.)

Another time I did a commercial for a Ram vehicle. I did a series of videos for a bank. An ad for a drink sweetened with honey. Once I was hired to wrangle bees for a television commercial for an athlete's-foot cream, which featured a fairly large pile of live bees atop the feet of a mannequin, later imposed onto the bare feet of a person supposedly suffering from a terrible bout of itchy, painful feet. Twice I took a box of live honey bees to a huge Tribeca photo studio for a shoot related to a cover story for *Time* magazine about colony collapse disorder, or the disappearance of the bees. Another time, my father, brother, and nephew, Patrick, and I showed up for a Honey Nut Cheerios commercial in which a lone honey bee was to buzz around a single flower. The commercial was shot indoors near huge, bright windows, and it was very difficult to get a honey bee to stay in orbit around just that one flower and not either head on a suicide mission toward the bright lights or toward the windowpane in a vain effort to escape. We tried spraying the flower with sugar water and even putting the bees briefly into the refrigerator to make them less active (not our finest moment, perhaps) until we noted an ASPCA representative on the set and decided against trying that again.

Way back in the 1990s, my father and brother, Mike, were hired by *Late Night with David Letterman* to accompany another beekeeper who was bringing in live bees to be used as part of a skit. The unnamed beekeeper transported the bees in a glass case that was poorly sealed, and the routine became much more dramatic than intended when the bees escaped. The mystery beekeeper promptly fled the Ed Sullivan Theater, and my father and brother were left to try to capture the hundreds of honey bees that were homing in on the powerful television lights, where they would scorch their wings and fall alive, but helpless, onto the stage. It wasn't ideal. A quarter century later, in the Ed Sullivan Theater again, Stephen Colbert from *The Late Show* borrowed one of my bee veils to do a skit in which he discussed having no apes in his apiary. (Get it?) Years later still, a swarm settled on a traffic signal just in front of the theater—it had come down from the aforementioned world's highest apiary, more than likely. We borrowed a ladder from the studio and took the bees home.

More recently I dumped live bees all over a stuntwoman for an episode of *The Blacklist* and enjoyed chatting with James Spader in between takes. At one point when we were reading through the script together, I apparently made a face in response to one of the lines.

"Andrew? What was that? Is there something you have to tell us?" Spader said in the manner a vigilant teacher might use to address a pupil caught talking to a friend during class.

I shifted my eyes to Allison, one of my helpers that day. She gave a soft shrug as if to say "This is your problem; I cannot help you." There were about twenty people crammed onto the set, which was outfitted with faux beehives and all sorts of beekeeping gear. I looked in Spader's direction and gently pointed out that the queen bee did not actually gather royal jelly—or whatever incorrect behavior the script had her or one of her hive companions engaging in. Almost immediately we changed gears, and before I knew it the two of us were sitting side by side poring over the minutiae of the script as it pertained to honey bees, two dozen crew members silent and looking none too happy.

The producers had to send my revisions back to California each time I made a correction. Spader was pleased, as he was as meticulous as his character on the show, and wanted everything to be perfectly accurate. The crew was clearly displeased, as the day dragged on for fourteen hours due in no small part to the script review and the wait for script approval from the higher-ups. I made no friends, aside from Spader, that day. The stuntwoman fancied herself something of an expert on bees and everything else, did not take instruction well, and among other departures from normative behavior when handling live bees, made the mistake of being too much in a hurry—not a good idea when it comes to *Apis mellifera*. She ended up being stung dozens of times. But they "got the shot," so the producers didn't balk.

The Blacklist, of course, was a high-end production. On the low-end side, a woman once tried very hard to convince me to cover her in honey bees as she did a tai chi exercise. She also wanted me to find a rooftop on which to perform this escapade. And she wanted me to volunteer my time since she would be "promoting honey bees." I can't remember what her motivation was. Lack of adequate prescription psychiatric medication, probably. I passed on that one. Another request I let fall by the wayside was from a bartender in Hell's Kitchen. She wanted me to gather bee venom, which she would, she imagined, spread on the rims of the glasses "to make the drinker's lips feel numb and tingly." Never mind anyone who might be allergic, I suppose. So beyond just maintaining my small charges in a beehive and stealing their honey, beekeeping in and around New York City has led me to some unusual situations with bees and humans. And I love it all.

The most memorable television venture was when I was hired to shoot a commercial for the state of New York, one of those "I ♥ New York" commercials that show people in the other forty-nine states how great a time they would have if only they'd come spend their time and money in the Big Apple. We moved ten beehives from our Connecticut farm onto the rooftop of a tall building on Beekman Street in the Financial District. The building, which had formerly housed a century-old law firm, had magnificent detail within and a commanding view of lower Manhattan from above. Since this was during the time that One World Trade Center was being built where the Twin Towers had fallen, the view was both impressive and poignant. Among other scenes, the commercial featured, for a brief moment, me lifting up a frame of honey bees from a beehive, smiling at them, and looking tremendously pleased to be on a New York City rooftop. My scene was so short that if you sneezed, you would have missed it.

But it was an amazing piece of work in the end: The soundtrack to the commercial was "Empire State of Mind" performed by Jay-Z and featuring Alicia Keys, and the narration was done by Robert De Niro. But the most interesting and fun part for me was that Spike Lee himself directed the spot. I've enjoyed Spike Lee joints since I worked as a projectionist at an alternative movie theater back in the 1980s. At that point, the small, independent SoNo Cinema in South Norwalk was the only option for seeing many of the foreign films and smaller releases of the day without heading into New York City. It was there, working as a projectionist, using an old-fashioned and even then outdated two-reel projector system and switching seamlessly between the two reels by listening for the bells and watching out for the cigarette burns in the right-hand corners, that I got to know and largely admire the work of Spike Lee.

This admiration didn't diminish upon meeting him. Anna Veccia was helping me again that day, as was my father, Norm. Spike, raised in Brooklyn but originally from the great state of

Georgia (whose official state insect, like Utah's, is the honey bee), called my father "sir," which placed him on my good side immediately. Spike has a film production company based in Fort Greene, Brooklyn, called Forty Acres and a Mule. He employs a huge crew. Anna was in heaven working among an ensemble of friendly, muscular men. She was so smitten by her surroundings that she didn't even feel it when she got stung on her cheek.

Spike himself was a bit smaller, older, and rounder than I remembered him from his days on screen as a pizza delivery boy in *Do the Right Thing*. He was also terrified of bees. As various preparatory tasks were performed by his underlings, we spent two hours together chatting about bees while he maintained a safe distance from them, and for extra measure and in spite of the heat, wore two hoodies and a bee veil. Maybe this was as a precaution in case any of the bees had seen either of his less impressive works *She Hate Me* or *Red Hook Summer* and wanted revenge.

Spike's questions were basic and genuinely curious ("Do bees sleep?" and "Is honey their vomit?"). Naturally, my questions on the subject of filmmaking would be just as elementary. But once we started shooting, it was clear that Spike was a master of his craft. The final product, the commercial, was great, and I learned something even more magical that day— all about the world of residual payments. That commercial ran around the country for months, and for every time it appeared on screen, I received a check. Those few hours rewarded me better than an average year of teaching.

In addition to the thrill of hobnobbing with the likes of Spike Lee and the craziness of fielding inquiries from tai-chi whack-jobs, occasionally there's the satisfaction of being involved with serious art. One day in 2014, when I was hawking honey at the summer farmers' market in Rockefeller Center, a nice woman approached me from the Museum of Modern Art to discuss the possibility of collaborating on an art project for the museum that would include live honey bees. The more we talked, the more interested I became. Part of the reason I was intrigued was because the project was familiar to me, in that the artist's studio had reached out to me about it via email a few years prior, in 2011.

Dear Andrew Coté,

I work with the visual artist and filmmaker Pierre Huyghe, a French artist based in New York. He is one of the most celebrated artists of his generation, having exhibited his work at the major museums and galleries worldwide.

The letter went on about what an incredible hero this guy was to the art world and how he was the best thing to come out of France since baguettes and Madame Curie.

Would you be willing to have a brief phone conversation with me as soon as possible? I realize you must be very busy but perhaps it's faster to have a phone call rather than answer email.

I was busy, but, of course, I did phone Melissa, the assistant who had reached out to me from Pierre's studio, and we had several conversations about how Pierre could succeed in his idea for incorporating live bees into his art. But soon, in what was and is a pattern, I could sense that the person on the other end of the line didn't so much want a few minutes' worth of advice as they did a full-blown complimentary consultation. Dope that I am, by the time I fully realized this, I had pretty much unloaded all of my ideas on the assistant via various chats and emails. When I broached the subject of establishing a paid consultation to continue, contact on their end immediately ceased. They later fully implemented my ideas and insights, giving no credit or payment whatsoever to me. I learned from the experience; now when people attempt to speak to me about special projects, particularly when attempting to utilize honey bees in unconventional ways, I'm pretty tight-lipped until we have some sort of written understanding. Sometimes people take this the wrong way. Though no one expects an attorney, a therapist, a plumber, or any other professional to work for free, it seems that people are often so charmed by honey bees that they forget that the individuals who keep those charming creatures need to make a living, too. If someone who keeps beehives wanders up to me at the Union Square Greenmarket, has a routine question about bees, and I am not too busy, I am more than happy to talk shop and give advice as I am able. But this was different.

My experience with MoMA was the polar opposite of that with Pierre's studio. Everyone I came into contact with from MoMA was professional, appreciative, and serious. My interaction with them all began when a wonderful woman named Lynda Zycherman sent an email to the NYCBA in 2014, about which I was naturally curious for a couple of reasons:

I am one of the Conservators for Sculpture at MoMA. I was hoping to get in contact with Andrew Coté, at the suggestion of Lou Sorkin, Entomologist at the American Museum of Natural History. I need to confer with a bee-keeper about a sculpture that includes live bees. Please contact me ASAP at the numbers below, or at this e-mail.

I remembered speaking on a panel with Dr. Sorkin a year or two prior at a Slow Food event, where we'd chatted over wine and cheese. I was happy that he remembered me at all, let alone that he thought enough of me to recommend me to the Museum of Modern Art. When Lynda Zycherman and I, along with her assistant, Ellen Moody, who is now an associate conservator at MoMA, met in September 2014, we did so at the Rockefeller Center Greenmarket, as arranged in our initial email discussions, since it was just a few blocks from the museum. It was a few days before Rosh Hashanah, the Jewish New Year, which is celebrated in part with dipping apples into honey to ensure a sweet new year. Lynda mentioned she would be out of the office during the High Holidays, which in her case meant about two weeks, and so I loaded both Lynda and Ellen up with various honeys for their respective apple dipping and challah munching, and we made plans to meet in a few weeks. It was an auspicious beginning.

And meet we did—and meet and meet and meet. Sometimes two of us, sometimes three, sometimes twenty of us at a large table filled with curators, security, public relations, attorneys (always attorneys), and, it seemed, someone from every department of the museum. I got a real taste of the bureaucracy and scope of a place like the Museum of Modern Art. I am in awe of the work that these folks do in housing and displaying the artwork, and educating the public as one of the largest and certainly most important museums of modern art on the planet. The museum—well, to describe MoMA is well above my pay grade so leave it at this: It is an inspiring museum, and though they are routinely changing the exhibits and bringing in new pieces, only about 3 percent of its collection is ever on display at any one time. So to have a place like this send some of their people to come and speak to me while I foisted honey on tourists made me feel very special.

MoMA contacted me based on a personal recommendation, but I am often the go-to New York City bee guy because I founded and am de facto head of the New York City Beekeepers Association, which means I'm highly accessible. Woody Allen said some variant of "80 percent of success in life is just showing up." When beekeeping needed to be legalized, I showed up. I was the one who started the NYCBA and spent time preparing the classes, teaching the courses, organizing the meetings, finding speakers, renting halls, answering thousands of emails, opening bank accounts, doing tax prep and paperwork, showing up to events, visiting schools to give talks, and everything else that goes along with leading an organization with a mission "to provide our members with a medium for sharing knowledge and mutual interest in beekeeping, and to educate and promote the benefits of beekeeping to the world, in a forum of friendship and fun, and to do so safely." Not at all on my own, I rush to say. There were many others who also dedicated and volunteered their time toward bee-centric goals: David Glick, Bettina Utz, Molly Conley, Paulo Anjou, Kelly York, Gerry Gomez Pearlberg, Vivian Wang. But after a few seasons, they largely came and went as life shuffled them to and fro. Jobs relocated some, children emerged from others, and priorities shifted from bee brood to their own brood.

In time, after a great many meetings and conferences and phone calls and insurance policies and site visits, there was an agreement that MoMA would purchase a statue from Pierre, I would ensure that the honey bees built their hive as specified by the artist's instructions, which were in reality just as per my 2011 instructions, and the museum would have the first exhibit containing live creatures in its nearly ninety-year history. I was chuffed. I was also amused and felt some satisfaction that the museum had hired me to curate the beehive on the sculpture of the artist who designed the beehive on my suggestions.

The head of the statue was a concrete copy of a 1930s Max Weber statue, and the body that of a woman who was lying down and propped up on an elbow. It was delivered in a wooden crate to our small, unprotected, totally unsecure farm in Connecticut. We lifted it off the delivery truck with our forklift and set it down, cracked it open, and tried very hard to be impressed. The departure from the Weber version was immediately visible in the form of sheets of honeycomb-imprinted plastic that obscured the reposed woman's head and face. The idea was to entice the honey bees to build an exposed nest there so that the head of the woman would become a seemingly feral colony of nesting bees. Not wanting to shine too much light on the magic, I won't explain exactly how this was done. I will say that honey bees in North America are not generally accustomed to or fond of building outside of a small, dark, dry, secure space, so this was slightly challenging. If we had been in the Himalayas, on the other hand, this might have been easier, as *Apis dorsata laboriosa*, native to that part of the world, enjoy building open-air homes with the comb exposed alfresco.

But there were stumbling blocks in accomplishing this in New England. In the process, I lost two colonies to absconding. Whereas a swarm is how a beehive splits from one into two colonies, each complete with its own queen, when bees abscond, 100 percent of them leave in search of greener pastures. I watched one group fly away and land on the eaves of the house next door, and a few hours later, mercifully, move again, to where I could more easily grab them. Another left without leaving any forwarding information or eyewitnesses. There is only so much bees will tolerate before they give flight or otherwise express their disapproval.

Which is partly why I decided to place two beehives atop the roof of the museum. I wanted to be sure that if things went wrong with the sculpture I would have a replacement colony—and a backup colony for the backup colony—ready for action the same day. To that end I reached out to my trusty friends the Pintos.

Dennis Pinto was born in Nairobi, Kenya. His wife, Joy, like me, is from Connecticut. They have two teenage children, Sasha and Tristan. Dennis and Joy run a family safari company, Micato Safaris, that covers much of the continent but specializes in East and Southern African safaris. The Pintos had heard about my work with the Samburu tribe in Kenya, and we hit it off. Once the elder Pinto child, Sasha, had peeked inside a beehive at a friend's school, she was hooked. Generally I don't take on youngsters, but these kids were so bright and eager that I couldn't say no.

When it came time to place two beehives atop MoMA, Sasha and I decided that we could not just slap two average beehives on the roof. "Even if no one sees them, they need to be special," Sasha implored. She was right. So I dreamed up the idea of building two hives that she then decorated to look like little New York City buildings. (I dreamed, Sasha built. It was a great arrangement!) One was reminiscent of a pair of townhouses and the other looked like a tenement, right down to the exterior fire escapes. Into the tenement hive we placed a strong colony of Italian honey bees, which quickly overpopulated to the point where bees were always hanging around on the tenement's stoop and fire escape. Had we hung laundry it would have looked perfectly like a nineteenth-century tenement from the Lower East Side. Unfortunately, we did not build our own works of art to withstand being out-of-doors. By the end of their three-month stay on MoMA's roof, the hive apartment buildings appeared dilapidated, not unlike a lot of city buildings in the 1970s. In both incarnations, first pristine and later ramshackle, they were spectacular-looking abodes for the bees.

Prior to and in preparation for the success of the New York installation, my father and I traveled to Los Angeles, where a different copy of the same Huyghe statue, called *Untilled*, was on display. We learned from the successes and difficulties the Los Angeles beekeeper had faced with the project. While there, we rented a yellow convertible with a black top, because, well, for obvious reasons. I also communicated with the beekeeper in Germany who had previously managed the same installation in Europe. Of course, I was in constant contact with Huyghe's extensive team of young French female assistants. Huyghe and I spoke over the telephone from time to time. Sometimes he was in France, sometimes Chile, sometimes elsewhere. We never met in person, but we both had pregnant partners at the same time and we diverted to that topic on occasion. It was a strange but ultimately beautiful collaboration.

As far as the main attraction—the sculpture—went, in the end I was fortunate enough to get the bees to follow the basic plan. Truth be told, my father came up with the idea that turned the tide for art's sake and encouraged the honey bees to build as we desired them to. Two months into the preparation of the sculpture I had a visit from a sampling of the assemblage of what seemed to be exclusively aesthetically favorable young women from Huyghe's studio. They were unapologetically late, and mostly stood around smoking and looking bored while criticizing the progress of the honey bees and probably the beekeeper.

Watching them brood and smolder while mumbling rebukes in French made me feel like I was watching a French film back in that projectionist booth at SoNo Cinema. Mostly, I gathered, because they needed to justify their forty-five-minute journey outside New York City, I trimmed, reshaped, and manipulated the comb while the assistants took photos between drags and sullen looks. They texted these (the photos, not the looks) to the artist. Soon we had the bee-headed woman looking as the artist desired, in the natural way he wished, through trimming and cutting.

I knew that eventually, the following month, the sculpture would have to be taken to MoMA. Usually costly works of art are transported by specialty art moving services. This work of art was now also home to about fifty thousand bees and counting. MoMA consequently decided that I should be the one to haul it to the museum for an installation at sunrise on a chosen day. I don't think this decision went all the way up the supply chain, and, in retrospect, it feels like it was reckless. But my father and I lifted the sculpture, apparently valued by some people at over a million dollars, with our secondhand dilapidated forklift into the bed of my aging Toyota Tundra, and strapped it in four times over. We had modified the cargo box to provide for airflow without allowing a direct breeze onto the bees. Satisfied that the sculpture and the bees were as safe as we could make them, we drove into the city.

I parked my truck on Amsterdam Avenue near where I was living at the time, and figured that most likely no international art thieves also endowed with extensive beekeeping skills would detect and have the ability to remove the piece during the ten hours it was parked there in front of a Chinese food takeaway shop. No bees buzzed around the vehicle, but there was netting over the entire bed of the truck just in case any hobos lurked, and the bees were well situated for the night.

Later some of the MoMA staff, particularly Margaret Ewing, a contemporary art specialist who was my liaison at the museum, nearly had a heart attack when I told her that I had not driven in that morning as they had assumed, but the previous evening. Their precious new work of art had sat all night on the mean streets of Manhattan. But I had come in the prior evening out of an abundance of caution, wanting the bees to be as calm as possible for the installation. Any disruption to their routine would impact their behavior, and bumping along in the stop-and-go rush-hour traffic of the Cross Bronx Expressway would almost guarantee a festival of stings at the unveiling. I thought that having the rough road behind them with a good chance to settle in overnight, followed by a short drive of a few blocks to their next abode, would put them in better spirits. Plus, I would be able to sleep a bit later and not worry about traffic jams or construction detours.

The morning that we were to install the sculpture was not difficult except insofar as my patience was concerned. I was amazed at how many people were involved. Though I had enjoyed free rein with the silent reposing woman for a few months, suddenly she found herself being handled by many men unknown to her. During transport I had a mild concern that the comb could have shifted, that the queen could be crushed, that perchance the country bees just didn't wish to become city bees and might abscond. I had invited several beekeeping friends who were also art lovers, as this was truly a special opportunity—fun for them and good for me to have bee backup in case there was an unforeseen issue.

An obviously pregnant Yuliana was there in her high heels and tight-fitting dress on her way to work near Rockefeller Center. (Note about Eastern European women: They can walk on winter sheets of ice, climb ladders, and ascend volcanoes in high heels, and make it all appear effortless.) Lastly my friend Jennifer accompanied me. We sat on the tailgate of my pickup with the bees buzzing in the box behind us and watched the teamsters, the MoMA staff, and dozens of others scramble here and there in preparation. Mostly we waited. We decided that in celebration of the day and as a nod to the artist (but mostly in deference to our hunger), Jenn, who grew up in Paris, and was therefore well qualified to choose croissants, would dash off to a nearby French bakery for chocolate and bread. We feasted while parked near the Fifty-fourth Street entrance to the museum.

After a number of hours and considerable ennui, the sculpture was installed; the bees were intact and pleased—even blasé—to have made it from a New England driveway to the center of Midtown Manhattan. After much scrutinizing over the precise angle at which to place the piece, where this rock should rest, how to adjust that tuft of ivy, whether to include this leaf or that one, we were, at long last, finished and able to take the screen covering from the bees and let them do whatever it was they'd collaborate to do. Luckily, that was just about nothing. They were released with a whimper and did not really stir until the dappled sunlight peeked through the branches of the European weeping beech trees that shaded them. So there it was. A huge hunk of cement in the shape of a woman with the head of an active beehive. It was, to bee-keepers, interesting and to civilians thrilling. The new and unconventional apiary was set up on a bed of ivy within ten feet of the museum's walking paths. I have to admit it was pretty exciting to be a part of it all.

The piece sat in the sculpture garden from June until late August. The hive was robust, the queen strong and prolific; I checked on the bees twice a day, partly to make sure that there were no signs of an impending swarm brewing, but mostly so I could place a screen cover over the entire thing at night to keep the bees tucked in for the evening. I would remove the screen in the morning. It was apparently in no one's job description at MoMA to get that close to the bees, and so I visited that sculpture garden fourteen times a week for months. This was not part of the original job description, but I enjoyed it, aside from the fact that I was unable to travel farther than a few hours from Manhattan for the entire summer due to this obligation. A huge benefit of having to go there very early in the morning prior to opening and again after closing meant that I had the place to myself. Though there is round-the-clock security, no one wanted to go near the bees, and once they knew who I was, security just waved me in and left me to my own devices.

And it was amazing. The sculpture garden has been on the same site between Fifty-third and Fifty-fourth streets and Fifth and Sixth avenues since 1939. According to Glenn Lowry, the museum's director, "the current garden was designed in 1953 by MoMA's director of architecture and design, Philip Johnson, and is dedicated to Abby Aldrich Rockefeller, one of the museum's founders." Her townhouse once stood on the site. "But in 2004, architect Yoshio Taniguchi's new buildings restored the transparency and continuity between the interior and exterior spaces." It is a gorgeous oasis in Midtown.

I met Glenn one day when I was waiting for the museum to close, and I was answering questions in the garden for curious art lovers turned bee enthusiasts. Glenn and I chatted for a long while, and when he walked away I said to Margaret, "That guy was nice."

"He's the director of the museum," she responded, probably in quiet terror the entire time that I would say something inappropriate, as I sometimes, though unintentionally, do.

* * *

Whenever placing a beehive anywhere it is important to consider not only the needs of the honey bees but also the perception of those honey bees by the humans who come into visual contact with them. People are not always well informed about bees and as often as not will allow their prejudices and fears to guide their feelings. That summer, some people expressed concern about bringing bees into the city at all. But even if this colony had not been gracing the sculpture garden, there would have been 258 or so types of bees flying around New York City, as there are right now, and that does not include the wasps. Bees live peacefully among New Yorkers, most types have been in the city for a good long while, and without them we would have significantly less pollination. I find that education about honey bees and their beneficial and docile nature wins over most worried civilians. The rest can elect to stay away from the hive, I say.

In regards to special considerations when installing *Untilled* in the sculpture garden, we wanted to establish a natural barrier that would prevent people from coming too close to the hive for the safety and protection of both the bees and visitors, yet not interrupt the viewing pleasure of John and Jane Q. Public. Bees need a bit of space to take off and land, and that space needs to be free of the more stouthearted art aficionados. As inquisitive as people are about art and bees—or in this case, bees as art—most possessed a level of self-preservation that kept them from getting too close to or too curious about the installation. Through the natural barriers that existed with the meeting of the stone patio and the greenery, and the constantly present vigilant security team members, things went smoothly and there was not one sting all summer.

The bees did very well in Manhattan and were seemingly quite pleased to be a part of the New York City art scene. Few beehives can boast being in such good company—works by Henri Matisse, Henry Moore, Pablo Picasso, and other great artists were the immediate neighbors to these tens of thousands of honey bees. They were enchanted by live classical music or jazz on Sunday evenings. They may well have been the most photographed colony of bees ever, with thousands of people from all over the world snapping shots of them seven days a week. Yoko Ono even stopped by to greet them. They were lucky bees indeed.

Despite this luxurious setting and the constant attention, the forager bees did just as their name implies they will do— they foraged for nectar, pollen, and water, and they did so traveling up to three miles from their home. Most, according to the orientation of their flight paths, headed to Central Park to enjoy the acres of bounty there. Many others made beelines for Park Avenue, where they feasted on the smorgasbord of nectar on offer there. Their honey may be the most complex, cosmopolitan honey ever concocted by the alchemy through which bees spin nectar into liquid gold. "There are always flowers for those who want to see them," that wild beast Matisse once said.

The installation got a lot of positive reviews and the press had fun with it, offering headlines like MOMA'S LATEST ART PIECE IS CAUSING A BUZZ, reporting that included "thousands of visitors have made a beeline for [the piece]," and comments like "She's bee-headed." Honey bees and beekeeping naturally lend themselves to puns and wordplay. The fact that we so easily anthropomorphize honey bees is because it is nearly impossible not to do so, and we have no good reason not to.

Having unfettered access to MoMA for the entire summer expanded my cultural horizons. I was able to halfway meet Yoko Ono when I ran into her. I wasn't running, but we did physically bump into each other, ever so gingerly, as she was in the basement looking at a collection of foreign versions of Martin Scorsese film posters. (Scorsese's 1973 *Mean Streets* was filmed, in part, in Old St. Pat's cemetery on Mulberry Street, where I had the seven beehives.) I was there due to a love of Scorsese films and a yearning for the uncrowded restrooms I knew were on that level.

I was struck by how absolutely tiny Yoko is. A wisp. She had on her trademark large dark glasses and a wide-brimmed hat. For a woman of some eighty-plus years she was very well put together, at once frail and a powerhouse. Someone from the gaggle of people she was with told her that I was the bee-keeper for the piece in the garden. She said something about bees being in danger, but was quickly swallowed up by her entourage and shuffled in a different direction. Brief as it was, it was thrilling for me to have brushed up against such a leg-end in the art community. Though I earlier gave the bees credit, Yoko had perhaps truly been the first to bring insects and art together at MoMA in 1971, the year I was sprung into the world. At that time, she had a one-woman show there that was marked by her releasing "flies onto the museum grounds. The public was invited to track them as they dispersed across the city." Art!

Yoko was around quite a bit that summer. She had another exhibition the same time the bees were featured, a revival of her one-woman show featuring art from 1960 to 1971. This included her well-known *White Chess Set*, which I was able to play, though it's unclear if I won or lost my match. Which was her point.

Every Friday, Yuliana and I would meet at MoMA and have drinks and something to eat at the Modern, one of Danny Meyer's restaurants, which faces the courtyard from its home on the first floor of the museum. Danny Meyer is a legend in the restaurant game, and his company even cooks for the staff cafeteria, providing gourmet restaurant-quality meals to the 750 MoMA employees at very low prices. I ate in the employee cafeteria several times per week, since regularly patronizing the Modern would have impoverished me. Danny is the master-mind behind the now global Shake Shack, and several other well-known restaurants. He and I have twice given talks on sus-tainability and locally and ethically purveyed food at his Union Square café. He is a friendly and generous man. So I was happy to sit on the patio of this high-end haunt, drink what-ever was put in front of me, and wait for my sixty-second task of putting the bees to bed for the evening.

Many evenings the museum hosted crowded live music events in the sculpture garden. Whenever that would happen, I would be called upon to place a large screen cover over the reposing bee-headed woman to ensure that the bees would not be riled by the commotion of the thousands of people in the courtyard. It occurred to me that hundreds of them would have panicked had they known how close they were to tens of thousands of bees. Of course, all parties were perfectly safe, even as the courtyard was filled with the sounds of live jazz or classical music. For nearly half a century, MoMA has hosted free concerts in the sculpture garden, featuring Juilliard School musicians and Jazz at Lincoln Center talent. Really, the bees that lived in the sculpture garden that summer were probably the most cultured ever to be found in the history of the world.

All good things come to an end. By late summer the exhibition was over. We gently vacuumed the bees off the comb with a special low-suction bee vac that safely removes them without doing any harm. Sasha and I reluctantly removed the townhouse and tenement beehives from MoMA's roof. We relocated those bees an appropriate distance of more than three miles to prevent the foragers from returning to the same area in search of their former home—all the way out to Rockaway Beach, in fact. The sculpture herself was taken away by teamsters and put into cold storage, which, unclothed as she is, may prove a hardship for the reclining lady. And with her departure, my time at the MoMA came to an end. Sasha and I took the bees from *Untilled* and relocated them to the apiary we had installed on the rooftop of the Rudolf Steiner School's high school on the Upper East Side, where Sasha was a student. All wrapped up, I bade farewell to my summer of enlightenment, returning again to my life as a philistine beekeeper.

JUNE

I love to see a swarm go off—if it is not mine, and, if mine must go, I want to be on hand to see the fun.

—JOHN BURROUGHS, *Locusts and Wild Honey*

Once upon a time our family sold honey and other products from our hives at local farmers' markets in Westport, Greenwich, Fairfield, and New Haven in our home state of Connecticut. The Westport Farmers' Market began in the back parking lot of the Westport Country Playhouse, a regional theater with a Broadway-caliber stage, which at the time shared the space with the Dressing Room, a farm-to-table restaurant run by Chef Michel Nischan and sponsored by his business partner, the actor and philanthropist Paul Newman. The restaurant was a huge hit with the patrons of the playhouse and supported the market in many ways, including the volume of food it purchased from the farmers.

The market was not large at the start, with only about a dozen local farmers. One vendor was Goat Boy Soap, founded by Lisa and Rick Agee. The Agees were motivated to begin keeping goats for their "severely allergic son, Bobby, who was able to tolerate goat milk and quickly became obsessed with all things goat." The Agees have used our family's honey and beeswax in their soaps for the past dozen years, and we could not ask for better partners. There were also organic vegetable farmers like Laura and Dave from Riverbank Farm out of Roxbury, Connecticut, and a butcher, a baker, and me, a (beeswax) candlestick maker. From a small startup market, it has grown into a massive weekly event, outgrowing its original space. The market has been masterfully run for the past decade or so by Lori Cochran.

In its heyday, the restaurant was always busy, and Newman was a regular fixture at the market and the restaurant. I saw him often and we spoke a few times. On one particular day, I was thrilled—initially—when Newman sat next to me at the small bar of his restaurant, where I was drinking boilermakers and waiting for a friend. He turned to me and smiled. My heart raced, my face probably flushed. In my head, I imagined per-mutations of the witty conversation to come, where I would talk about bees and he would be riveted, and he would share secrets from his days on the sets of some of the most iconic movies ever made. We would discuss his food business and my honey busi-ness, perhaps partnering together in the most charming honey enterprise ever fathomed by mankind. Now was my moment.

Then Newman asked me a question about the football game that was playing on the bar's television. I froze. Not only was I woefully ignorant about football, I was about to disappoint Butch Cassidy. I probably knew as much about football as Newman knew about selective queen breeding. I cursed myself for being unable to respond. Newman turned back to the screen, smiled, and charmingly ignored my obvious distress. I cleared my throat to speak, searching for something I could say about the game that was obvious and not likely to be wrong. Before I could open my mouth to place my foot in it, divine intervention, or the fast pace of the game, saved me. Something happened—a basket or a home run, I am not sure—but it caused Newman to pump his fist once in the air and exclaim some sort of celebratory line. I exalted in kind, saved from humiliation, and downed the contents of my glass.

Some years later, after Paul Newman's beautiful blue eyes had closed forever, I sat in the restaurant's mostly empty parking lot finishing a phone call before I delivered a bucket of honey to the kitchen. Honey is heavy, so I backed up toward the kitchen door, grabbed the sixty-pound pail from the rear of my truck, and placed it on the ground so I could park without blocking the service entrance. Between the rear axle and my front door, a man in a nicely fitted dark blue suit appeared seemingly out of nowhere, followed by two other fit, nicely attired men. They expertly triangulated me against the side of my vehicle. "What's in the bucket, sir?" one asked with restrained politeness. It was clear to me that these were not private security guys for some Hollywood big shot enjoying farm-to-table cuisine. These guys were Secret Service.

"Honey," I answered with rare sarcasm-free candor (and hoped he wouldn't think that's how I was addressing him). I stood with my palms facing the men, chest level, fingers spread. One of the agents picked up the bucket and carried it away from the restaurant as if it contained nothing more than feathers. Impressive. Intimidating. My initial reaction was to survey my brain for the transgressions that could lead to such a visit, but I quickly realized the situation.

"So," I started, "only two Town Cars here. You are . . ." I trailed off and started again. "So it must be . . . Bill Clinton inside there?" I continued, smiling. Clinton was no longer president, but he lived relatively close by. The looks on their faces should have confirmed that I was right, but these guys don't let their faces show anything. And let's be clear, guessing correctly didn't win me a prize. Rather, the remaining two agents took a step closer. PSA: Secret Service agents do not respond well to humor or attempts at charm.

Luckily, Chef Nischan came out smiling, walking briskly toward us. "He's okay! He's okay! That's our beekeeper, Andrew!" The agents glared for a moment, then, appearing almost disappointed, backed off. One laughed—sort of. It was more a hint of the idea of a smile, and for just a brief second. Chef shook my hand and told me that, yes, the restaurant was closed, and he was having a lunch with Paul Newman's widow, Joanne Woodward, and President Bill Clinton. "He's just finishing up some cornbread with your buckwheat honey drizzled over it right now!" Chef told me. How befitting that the Arkansas native was enjoying cornbread, and what an honor to have it drizzled with our honey.

One of the agents interrupted my moment of joy, his own mirth clearly over. "Do you need anything else, sir?" It was his way of saying "Get out of here."

Ignoring the question, Chef asked me, "Would you like to meet him?"

"Sure. Yeah. That'd be great."

"Wait here."

So I parked my truck, walked back, and waited. The well-dressed armed guards also waited. In about ten minutes, while he was escorted out to his waiting vehicle, the former leader of the free world came out and spent just over a minute talking to honey monger me. He is much taller than I expected. He's a controversial figure, no doubt, but there's no denying that this man emanates a palpable charisma, one I had never experienced in my life until that moment. I had heard—but until then I didn't believe—that Clinton makes the person to whom he's speaking feel as if he genuinely cares about him or her. "I loved that honey," he told me, and I honestly think he meant it.

Back when I was a full-time professor and a part-time beekeeper, I was able to handle doing one or two farmers' markets per week without neglecting any of my other responsibilities. It did keep me busy, though I had loads of help, of course. My father helped with the bees, as did my brother Mike and my nephew, Patrick, and all helped with the markets from time to time. My mother would sometimes bottle the honey, and my nephew often did markets on his own or with my mother. Once upon a time our family sold my father's honey under the label "Norm's Other Honey"—the first honey being, of course, his wife; my parents have been friends since 1955, a couple since 1960, and married since 1966. I liked to draw cartoons when I was young, so I designed a label for him that was a caricature of his head on the body of a honey bee. The problem was, at the tender age I inked the drawing, I erred and drew the body of a worker bee, so my father's head on the body of a female bee adorned his label for decades. He didn't want to make me feel bad about it, so he never told me; only in my late teenage years did I notice, and it was a source of small amusement to us and to fellow beekeepers.

That label remained until one day we were approached to sell our honey in a small locally owned group of supermarkets, and we needed a label with a UPC code. The store, a family-owned chain of three called Caraluzzi's, thought the label with the human head on the body of a bee a bit "strange," according to Steve Caraluzzi. So my father suggested that since I was now seemingly the heir apparent to our little honey empire—my brother was busy working full-time as a police officer and raising two children, and could only do so much in regard to the bees—that we might as well go with a new label. Thus Andrew's Honey reared its head (admittedly we get no points for creative names). My mother dug up a cartoon of a beekeeper holding a frame of bees I had drawn as a ten-year-old boy, and voilà, my own ridiculous label was born. Originally we printed the labels on regular paper and glued them by hand to the bottles. Later we graduated to sticky paper. Only recently have we put on big-boy pants, and now our labels are printed by an actual printing company.

We have always been content with our rather small footprint in the world of bees, which is fortunate, as our finite production means we can never become a big commercial endeavor. We have a constant flow of requests from markets and retailers who ask to carry our products, but we simply do not have the number of beehives necessary to produce the quantity that would be needed, and we cannot manage as many as would be required. Our beehives have never numbered beyond the hundreds, and though we produce tons of honey, we still fall far short of overall demand.

Interestingly, this resembles the gap between the global demand for honey and the actual amount of honey produced worldwide. For this very reason there is a brisk world market of fake honey. In fact, fraudulent honey is a billion-dollar industry; the volume of fake honey sold globally is exceeded only by the volume of fake fish and fake olive oil, also billion-dollar industries. Produced primarily in Chinese factories, the bastardized honey is usually one part honey and several parts cane sugar, rice syrup, or corn syrup. It is labeled as pure honey and then sold all over the world as the real thing, though it is a far cry from it.

If, for instance, we made and sold pies, we could probably buy the appropriate ingredients and move into a large commercial kitchen and amp up production and theoretically sell millions of pies, with purchased ingredients. But when it comes to our finite supply of authentic honey, we are limited to what we can harangue the bees to produce. Being that we live in the Northeast, the season is shorter than if we lived in Florida or California, and accordingly our yields are lower. Of course, there are large commercial beekeepers and honey producers even in upstate New York and Canada, so yes, it is possible to make a bigger go of it.

But we like our lifestyles, and where we live, and we have always been pragmatic and simple, perhaps even unambitious in the eyes of some. We never chased much after riches or made monetary rewards a priority. If we wanted to do so, we made a real mistake being involved in any sort of agricultural pursuit. Instead, we are content with and grateful for our share of the liquid gold our four-winged angels provide, and the satisfaction that comes with the labor invested on our end. Still, every now and then, when I am standing at a market in early February on ground that is as cold as ice and there is a gentle hint of dampness in the air accompanied by a strong steady winter gust on my face, I wonder, if it would have been so bad, if instead of being beekeepers, my forebears had started a bank or a law firm.

One day in 2006 my father and I went to visit one of Manhattan's farmers' markets. There was a fellow there named David Graves, who I had heard had beehives atop a building in Manhattan—this was during the Bitter Years. At the market we admired his display of jams and honey and maple syrup, and the photo on his display table of him invading his rooftop beehives while dressed in a bear outfit. Needless to say, we liked him right away. With a thick head of white hair and a very dry sense of humor, he talked to us about his rooftop adventures with the bees. Worth noting was that David drove all the way from Massachusetts to attend these NYC markets—five times a week, no less. On the train home that day my father said to me, "If that guy is driving four or five hours each way, he must be going home with that van full of cash." That's when we sent in an application for the biggest of the New York City farmers' markets, the Greenmarket. Maybe chasing after a little bit of riches wasn't a lousy idea after all.

Under the umbrella of GrowNYC, the Greenmarket, which is the name for the organization that runs nearly fifty farmers' markets throughout the five boroughs, was founded in 1976 in order to both promote regional agriculture (thereby supporting local farmers) and give New Yorkers access to the freshest possible local produce. The crown jewel of the New York City Greenmarket is the one at Union Square between Fourteenth and Seventeenth streets, held four days a week. Upwards of sixty thousand people attend on some days in order to grab locally grown produce and more.

"As Greenmarket's flagship market," the official description reads, "the seasonal bounty is unparalleled, with hundreds of varieties to choose from during any given season. From just-picked fresh fruits and vegetables, to heritage meats and award-winning farmstead cheeses, artisan breads, jams, pickles, a profusion of cut flowers and plants, wine, ciders, maple syrup and much more." We were part of the "and much more."

After a tremendous amount of paperwork filled out mostly by my mother, who is the bookkeeper to the beekeepers, we were approved and accepted into the system. We began by selling our honey and hive products in New York City at the now-defunct Grand and Essex streets market. It was a small and never terribly successful Sunday market (hence its ultimate demise), but we were thrilled to be there.

On the first day I set up, I arrived very early, around dawn, and found the neighborhood quiet. I had reached the market just ahead of the sun via the Williamsburg Bridge from Brooklyn. Some of the streets, like Norfolk and Suffolk, were cobblestone south of Delancey. There was graffiti in many places, trash everywhere, and several drunken people staggering home, or staggering somewhere. Soon the morning light on the Lower East Side began to bestow more color upon the area, and the pickle shop, bakery, and bagel shops across the street began to stir with people. The area was a blend of

Orthodox Jews and Chinese immigrants, with a large helping of Dominicans. The Orthodox Jews would not even taste the honey until I got a letter from one of the local rabbis affirming it as kosher.

The corner of Grand and Essex was an introduction to the street markets of New York City. For years after that I worked the market at Tompkins Square Park. I also set up shop in Forest Hills, Queens, and sometimes uptown on Ninety-seventh Street on the Upper West Side. One year I hawked at the market in Bay Ridge, Brooklyn, and also from time to time in Fort Greene. I still participate in the summer market held in Rockefeller Center, where my booth is a few feet away from the spot where the famous Christmas tree is set up in November.

All of these markets have, or had, their own blends of ethnicities and appeal, but none of them had as strong a lure as the Union Square market. And not just for the sales, though those are obviously better. But because the number of people, the celebrity spotting, and the sheer energy of the place make selling at Union Square a fun way to earn part of a living.

After ten years of applying and waiting and hoping, one year I was finally accepted into the Union Square Greenmarket on Saturdays. This is perhaps the largest producer-only farmers' market in the country. I was not on a waiting list; no list exists. Hopefuls apply each year and cross their fingers, and as needs arise and vacancies open up, they are filled by lucky producers. We were fortunate enough to get a spot when another beekeeper, Walter Bauer, retired, and we were selected from among many other eager and suitable candidates. We had already been selling our honey there on Wednesdays for some years, but to be there on Saturdays was life changing.

Now we have been able not only to showcase our products to a large audience and interact directly with a cross section of clientele that we would never have known otherwise, but also to steadily grow our business by reinvesting in it. Mostly, I have grown the New York City arm of our trade by installing bee-hives all over the five boroughs. Within a few years of our acceptance into the Union Square Greenmarket, I had bees atop the Waldorf Astoria, several other hotels and restaurants, in historic cemeteries, on churches and synagogues, in community gardens, atop schools, and on private residences. I even installed three beehives at the United Nations on international territory. The bees have been generally well received by the public, and I enjoy working with them in the urban setting and harvesting their honey to sell at the market.

Being in the big city, working bees and selling honey, has been good to me. But every June, I receive regular alerts about swarms of bees making trouble, and I drop everything to tend to them.

My pal Tim O'Neal, also a beekeeper, once said, "Trying to get bees not to swarm is like trying to get teenagers not to have sex." It's an entertaining observation, and in Tim's case, no doubt purely speculative, and I disagree. Hives of honey bees can certainly be inspected regularly and managed in such a way as to prevent swarming in the majority of cases. This, of course, does not always happen, and I will sheepishly admit that my bees do on occasion swarm. However, it is highly unusual for my city bees to swarm, because I check them obsessively to prevent the kinds of problems that are inherent with an urban swarm. It took a great deal of effort on behalf of a great many people to bring beekeeping back into the legal realm, and I want to be a force in word and deed to keep bee-keeping safe, responsible, educational, and fun. City swarms, while often providing entertaining stories, are not exactly posi-tive press. A swarm will also cut back on the productivity of the hive, and I want urban hives to make honey.

Swarms increase as the weather gets warmer, and I often assist the New York Police Department in corralling the quiv-ering football-sized lumps. This is how I met Tony Bees. Born in East New York, Brooklyn, the son of immigrants from Crete, Tony settled in Rego Park, Queens, where he lives and keeps beehives in his small backyard. Many years ago, the NYPD called me up to help them collect a swarm of bees from a tree in front of a clothing store in Brooklyn—the lieu-tenant on the scene happened to be from Westport, Connecti-cut, and he knew of me from the farmers' market there, and somehow the department had gotten hold of me. I was in Connecticut at the time, but I drove all the way down to the Bed-Stuy area of Brooklyn to help out. That's when I met Tony, who, as a bee enthusiast himself, had come out with his departmental colleagues that day, though he had no equip-ment with him at the time. Tony watched as my father, my friend Mio Shindo, and I captured the swarm with ease. We became fast friends, and for years captured swarms together and exchanged beekeeping advice. Tony even sometimes bought buckets of honey from me and sold it under his label.

Mio, visiting from Japan, was not a beekeeper except when she stayed with me. I met her in Turkey in 2000 when she was a high school student who had gone on her own to

Turkey to do earthquake relief work. We stayed in touch and she occasionally spent summers with me working at the farm. Later she graduated into swarm captures and feral hive rescues, and has since worked bees with me on BWB missions to Haiti and Tanzania. Now she works for the United Nations in places like Sarajevo and Lagos in a non-bee-related capacity, working with refugee populations—which, in a sense, swarms are as well.

My friend Jon Huang is a graphics editor for *The New York Times* and a beekeeper. He and his wife, Megan, generously offered their apiary in Manhattan as a place where some of the NYCBA apprentices could get their hands dirty (sticky?) with a different set of bees. When it came time for them to move their growing family across the Hudson to Jersey City, some volunteers from the NYCBA helped them screen in and carry their two strapping beehives down from the fourth floor walk-up where they lived. The NYCBA tries to cultivate the sort of congeniality found in the hive among its members and participants. Sometimes we get it right.

Jon texted me right away when he spied a swarm nestled on the limb of a tree on West Seventy-second Street one June afternoon. I was not too far away with my beekeeping friend Molly Conley, an Iowan transplant to New York, when I got Jon's message. The swarm was a bit high up for me to grab without my truck and ladder, but as I had a good parking spot, which is worth quite a bit to New Yorkers with vehicles, I didn't at that moment want to move the truck and lose my precious parking space just for the sake of three or four pounds of bees. Plus it gave me an excuse to phone my friend Tony and let him get one of his much-coveted overtime gigs, which was something I often assisted him in getting, at his insistent requests. So I called Tony, and we waited. And waited. Since Tony's overtime clock started when he started driving, he was not always in a hurry to reach his destination. Eventually he arrived, and though the swarm was benign and mostly out of public view, Tony had alerted the Twentieth Precinct to its presence. Only he must have presented it to them with as much hysteria as if a terrified toddler had been spotted dangling from the limb of that tree. I counted fifteen officers at one point and at least half a dozen vehicles. This was unusual, and there was no need for all of the backup. Two of the officers initially tried to hustle us away from the bees, until they spotted our beekeeping regalia and we were accepted as part of the operation.

Tony had also played his usual card and made some other calls—to the press. It was no accident that by the time he arrived in his vehicle there was a small swarm of reporters on the ground as close as they could get to the swarm in the tree. Still, there were not enough there to satisfy the Greek. Tony chatted with me, Molly, and Jon as we stood on the sidewalk waiting for him to begin the magic show.

"What are you waiting for?" I asked, grinning, knowing the answer. We all knew.

"Just a few more minutes." He smiled like a Cheshire Cat and winked. Tony was wearing a T-shirt that he had specially made that had his Hotmail email address across the front so people could contact him. "It shows up in the photos!" Tony told us, tapping his finger to his own head and smirking, letting us know he had thought of everything.

About forty-five minutes after his arrival, the size of the press pool seemed to meet Tony's satisfaction, and he ascended the short distance from the ground to the thin branch in a bucket truck.

Generally when I get wind of a swarm, I want to catch it myself. There is simply nothing about beekeeping that I enjoy more than capturing swarms. Particularly easy-to-grab swarms such as this one, which was drooping in one mass from the branch, not too high, easy to dislodge and carry off to a hive under bluer skies. Also, it is best to grab a swarm as quickly as possible. While they hang there, hundreds of scouts are out and about seeking a new home, and at any instant the swarm could come to a collective decision to depart. They might alight to a higher branch or move half a mile away. There's no telling. So securing them as quickly as possible is important.

Molly, Jon, and I stood on the sidewalk and cracked jokes to one another as we watched Tony in action. The show—or showboating—commenced. First Tony, who wasn't yet wearing a veil or gloves, leaned in to the swarm. He waited a moment to make sure that all of the many photographers were poised. Then he made his signature move of slowly plunging his bare hand into the thick of the swarm, an action that always produced a cacophony of *oohs* and *aahs* from the uninitiated, and some dramatic photos to boot. At this point his facial expression was stern and determined, as if he were trying hard to either void or retain his bowels.

Tony routinely claimed to neophytes that the immersion of his hand into the swarm got the bees used to his scent so that he could better capture them without incident. For the beekeepers watching, the giggle factor gets pretty high when hearing this explanation, but it's all supposed to be fun, so we don't spoil the entertainment. Capturing swarms is, after all, a simple procedure, so why ruin the spectacle with a few facts?

There have been more swarms than I could ever recall in New York City. There are dozens every year. But to actually see a swarm in flight is a rare and awe-inspiring thing. Rudolf Steiner said, "If you now look at a swarm of bees, it is, to be sure, visible, but it really looks like the soul of a human being, a soul that is forced to leave its body." And the Austrian esotericist was right. What he didn't mention is that it can be as loud as a buzz saw, and the bees may spread out over several cubic meters of airborne space. As for capturing settled swarms, one rare day Detective Dan Higgins and I captured five in five hours—our personal record while working together.

Years ago, I had a call about a swarm in the East Village. The bees had departed from a hive kept atop a walk-up building on East Ninth Street. I alerted a few other beekeepers, and we all headed to the spot. Then I got a follow-up call that the swarm had moved on. I asked the beekeeper to look carefully, in case the swarm had simply moved to a higher branch or another nearby tree. No dice. It was a warm day and perhaps the scout bees had found a suitable domicile elsewhere. C'est la vie.

Still, I was already en route, so I continued to the scene anyway. Sure enough, the swarm was there, just higher up in the tree, now about twenty-five feet from the ground. When I arrived, I realized that the person who'd called me had been in one of my beekeeping courses. This woman had not inspected her hive even once that year, and the swarm was, naturally, from her own rooftop hive. But it was too high to be reached by the ladders that we had available.

NYCBA member Paulo Anjou, a native of Monkey's Eyebrow, Kentucky, and recently anointed beekeeper, showed up that day on East Ninth Street to offer his help. So did beekeeper Ezra Hug, whose love for honey bees extended to giving his daughter the middle name Bee. Our biggest problem that afternoon was that we could not reach the swarm. Normal people would have left it alone and gone home. We hatched a scheme to stop a truck, offer them some money to park under the swarm, bring a ladder atop the truck, and try to reach the swarm that way. Soon we had flagged down a Heineken truck and perched a ladder on its roof. Passersby gawked. We clambered onto the truck. Even with the ladder bringing us to greater heights, the swarm was out of reach, though the branch on which it dangled could be gripped. A few minutes passed while I evaluated the options. More passersby stopped and pulled out phones to snap photos. Two people in bee veils on a big green Heineken truck. "Even in the edgy East Village, this was an unusual sight," recalled Vivian Wang, another newly consecrated beekeeper witness to the melee.

Paulo and I perched on the roof of the tall truck. Paulo held the ladder; Vivian held her breath. I scaled the ladder, stood atop the uppermost point of it, and shook the branch hard to dislodge the bees. The swarm plopped right into the bucket I was holding. Well, most of it did. Spectators hollered, "There's bees all over you!" Shaking the branch one more time to get the remaining bees that had regrouped, I tuned out the yells, unperturbed by the hat and cape of winged creatures I was wearing. Part of the job of working bees in New York is dealing with the public. Many on the street fancies him- or herself an instant expert, and offers unsolicited advice about— or, more often, criticism of—whatever tactic they might see the beekeeper employing. One has to become immune to their voices and tune them out, especially when working up high and risking safety. Even though some voices included well-intended advice, I needed to keep my focus on the task at hand.

I climbed back down onto the roof of the truck. Paulo and I brushed my cloak of bees into the container as best we could, and then we both climbed down from the vehicle. At first we tried to shake hands with the Heineken truck drivers, but seeing the cloud of bees around us they demurred, smiled, and drove off. There were still bees all over Paulo and me. A few landed on the crotch of a passerby's pants, which I decided not to point out, thinking there would be an unhappy ending were he to panic and swat at them. It was not the cleanest swarm capture we had ever made, but we had the majority of the colony safely in our screened bucket. Appropriately, we celebrated with Heinekens, but we had to buy them from a nearby bodega since the truck had already sped far away from the haze of bees that had blanketed it.

We find out about swarms in a variety of ways. Some swarms we grab on our own when we're informed directly via our Swarm Hotline on the NYCBA website. Sometimes the call comes from someone at the police department, like Detective Dan Higgins, who took over bee duties for the NYPD after Tony retired. Dan and I had met several years prior at one of the farmers' markets and talked bees. Then he dropped out of sight for a while to successfully battle cancer. He returned to the force a bit skinnier and with a new zest for life—and for bees.

I started working with Dan the second Sunday in June on the day of the National Puerto Rican Day Parade. One of the largest parades in the city, it draws over two million people to its gyrating route up Fifth Avenue from Forty-fourth Street to Eighty-sixth Street. The parade has a well-earned reputation for rowdiness. Naturally, a swarm of bees would be disruptive to the already raucous proceedings and incite further may-hem. Or maybe liven things up. But we couldn't take any chances.

For the second time in two years, a considerable-sized swarm of bees had landed on the corner of Fifth Avenue and Fifty-seventh Street, this one on a pedestrian traffic signal just in front of a Louis Vuitton store, and just a few steps from Trump Tower. The bees were poised there, tens of thousands strong, silently obeying the bright red illuminated hand and its message to STOP. Apparently this intersection was prime real estate for swarms. The one the year before had been catercor-ner to this in front of the BVLGARI shop, and Tony and I had grabbed it while standing on top of an NYPD Emergency Services vehicle. Clearly the bees have high-end shopping on their minds with these four corners housing BVLGARI, Louis Vuitton, Bergdorf Goodman, and Tiffany & Co.

There have been so many interesting swarms in so many odd places in New York City they could perhaps fill a book on their own. Swarms have materialized on the deck of the mighty USS *Intrepid*—twenty thousand fliers attacked the decommissioned vessel one late summer morning; on the umbrella of a hot dog vendor in Times Square; on a United States Postal Service mailbox downtown (air mail?); on the underside of the Brooklyn Bridge; on a streetlight forty feet above the street in Chinatown (perhaps looking to buy knock-off Louis Vuitton bags, jealous of their uptown swarming sisters, Yuliana suggested); on the statue of a lion in front of a house in Queens. And lest one believe swarms are limited to intimidating the population of New York City, think again. Swarms of bees have grounded planes at several major commercial and even military airports. Swarms have interrupted baseball games in San Diego and New York. A swarm of honey bees even attended Muhammad Ali's funeral, perhaps to say goodbye to he who claimed to "float like a butterfly and sting like a bee." A swarm of bees in Gurgaon, India, caused enough mayhem at a local police precinct that a detainee was able to escape. One swarm even landed on a man in England—on his head and face—and left him helpless for quite some time until they decided his head was not quite empty enough to make for a good nest, and so they flittered off again. But one swarm I will never forget was atop an empty building on a roof that is used only once per year, but is one of the most-viewed roofs in the world.

Starting in 1903, the building now known as One Times Square was built right in the center of Times Square, back when that little plot of land was still called Longacre Square. Despite the shift in names from one square to another, the area isn't and never was a square. It is more of a bow tie shape, consisting of two big triangles, not one square. Whatever the configuration of the streets, the building presently at the center of it all is named for its first owner, *The New York Times*. When she was still a relatively young thing of 52 (she is now pushing 170), the Gray Lady was churning out copy right there in midtown Manhattan.

The *Times* building went on to serve multiple purposes for the next century plus. Almost from the start, it used its position to monetize advertising space. In 1904, the first electronic signage went up on the sides of the building. Soon the signs became works of art, and people would come just to view them. By 1917, Camel cigarettes had installed a long-standing advertisement of a sailor puffing away on a cigarette, with a hole for his mouth that spewed steam, replicating smoke. Wrigley's chewing gum, Studebaker Wagons, and, famously, Coca-Cola, have all advertised on the sides of One Times Square.

At some point in the 1990s it was decided that advertising along the sides of the building would simultaneously generate more income than renting out the space inside and eliminate the problems inherent with tenants. Thus one day, other than a retail space on the ground floor, all tenancy ceased and the shell of the building was dedicated strictly to bright, flashing advertisements to the tune of hundreds of millions of dollars per year for the owners.

The single most recognized aspect of this huge flashing billboard-of-a-building is the top of it: Since 1907, every year save a couple during the blackouts of the Second World War, the great ball drop has been a part of ringing in the new year in New York City and, courtesy of television and live streaming, just about the entire world. At present, the nearly three thousand Waterford Crystal triangles that make up the ball, lighted by more than thirty thousand LEDs, hang far above heads atop One Times Square twelve months a year. The ball makes its illuminated sixty-second descent just the one time annually. And it was on the very top of that building that we were called to deal with a massive swarm of bees one June day in 2017.

Hannah Sng Baek and Gus Lodise were with me working some rooftop beehives down on Warren Street in Tribeca when I got a text about a swarm around Times Square. There had been a few in that area, and though it was swarm season and I loved capturing swarms, I had work to do, checking my own hives for swarm cells and production. I also had an appointment with a German journalist at a coffee shop and did not want to break the appointment based purely on the speculation of a swarm "in the area." So we as a trio walked a block to meet the reporter for the German magazine, but I was distracted by another notification about this swarm.

"We shouldn't bother," Hannah said. Usually she would have been right about an unverified swarm. We did not have all of the correct equipment with us. The traffic and parking would be next to impossible. The person who notified me had not provided a photo, which generally I require before setting off to a supposed swarm because people tend to exaggerate or just get it wrong. But there was an urgency to this alert, and there were multiple messages. So I made a phone call and asked the questions that I ask, which are phrased to garner responses that guide me in verifying whether we are dealing with a swarm of honey bees or with some other issue, like a nest of wasps, or a few wayward bees attracted to a melted

Popsicle puddle. Disconnecting the call I said, "Let's check it out." And off we went.

Hannah first approached my stand at Union Square in November of her freshman year at New York University, having read about me in a book by Robin Shulman entitled *Eat the City*. It was assigned to all NYU freshmen for some years, during which I had a steady stream of smiling young college students stopping by the Union Square Greenmarket asking "Are you Andrew?"

Hannah soon demonstrated herself to be an asset to my little operation. She had been interested in honey bees since she was seven. At that time, she made a poster about bees and planned to be a beekeeper when she grew up. So reading that book and then meeting me was, in her own words, "like meeting my hero!" With that kind of ego stroking, how could I not keep her around? Hannah has now worked with me for years, at the market and on rooftop beehives. I know her parents and sister, I attended her college graduation, and she has visited and spent time with my Japanese "family" in Kyoto. On her own, she has gone off to work bees in the south of France; in central Russia an hour outside of Ufa, in an ethnically Turkish Muslim region; and has reared queens in Finland at the country's largest apiary with five hundred beehives. So she is a worker bee dedicated to the craft. And she is not afraid of heights, which on that day was a distinct plus.

I decided to leave my pickup truck downtown and risk a ticket. One uptown subway ride later, Hannah, Gus, and I emerged in the thick of Times Square. We were surrounded by people dressed as superheroes, people in red hats and vests trying to sell us bus tours, and multitudes of tourists. We were told originally that the swarm of bees was terrorizing a police substation at ground level, but they had left that site much earlier. The information I now had was that the swarm had taken hold up on the roof near where the New Year's ball dropped. We entered One Times Square to speak with the property manager.

We were wearing our beekeeping gear, since we were beekeepers beekeeping, and as a trio we appeared to be perhaps hazmat workers or extras from *Silkwood*. This appearance helps a great deal when trying to gain access to a security-intense building or otherwise hard-to-enter area; seeing someone all garbed like that lends a sense of urgency to the situation at hand and generally speeds up access. Gus was fifteen years old, the son of friends, and I did not want his youth to slow us down, should the management deem him too much a liability, so I told him to put the veil over his face to hide his juvenescence. Gus was spending a week helping me at farmers' markets and in other beekeeping work like digging holes and bottling honey. Or, put another way, he was learning about the hard slog that is real beekeeping. It was a great arrangement for me. I found myself liking him tremendously—he was smart, polite, not too talkative—and since I was also fond of his parents, I decided I needed to return him intact. I made a mental note not to allow him too close to the edge of the roof.

So there we stood, waiting for the building manager in our sweaty beekeeping gear, watching the tourists lined up to buy bottles of water and selfie sticks. They regarded us with open curiosity. This, along with their girth, and brightly colored fanny packs, was one of the hundred ways we knew they were tourists. New Yorkers are unfazed by our beekeeping gear.

They won't give us a passing glance. Most New Yorkers would step over a man on the sidewalk with a knife sticking out of his chest without lifting their eyes from their phones, except maybe to take a video of him. Also, New Yorkers avoid Times Square as if it will give them herpes, which, before Giuliani cleaned it up in the late 1990s, it clearly would have.

It is worth noting that, strictly speaking, this building's management had not reached out to us. The manager knew that there was an issue, but did not know exactly what it was or whom to call. This is common. Time is of the essence with regard to swarms, and because people in general just do not always know whom to contact, by the time the appropriate party is found the swarm has often moved on. Frequently this move exacerbates the problem, as the colony may find its way into a crevice in the affected building and set up shop, bringing greater and more difficult-to-resolve problems later.

In this case, though the management had not gotten hold of us, they had been trying to get hold of someone, anyone, to assist with the problem. Upon seeing the three of us all suited up, the manager took us straight up to the roof with nary a question asked. On the elevator ride up, I gathered that he did this not out of concern for the welfare of the bees, but simply because the building's union employees would not go anywhere near the roof due to the bees' presence. The advertising signs could not be adjusted, fixed, or otherwise physically handled, which meant that serious income was being lost, or soon would be. So the bees needed to go.

After ascending seventeen floors, we exited the elevator to see that the inside of the building was empty—there was no furniture, the walls were unfinished, pipes were exposed, and wires dangled from the ceiling. We were marched out onto the roof, where we were able to see the backs of the illuminated signs and a hint of the sky and buildings across the street. But there was no swarm to be seen. The manager stood in the doorway (not wanting to venture out, lest he be carried off by the bees, I suppose) and pointed to where the swarm of bees had settled earlier. By this point, there was only a bit of wax foundation there, evidence that the bees had begun to build a small amount of comb. There certainly had been honey bees there.

The roof felt surprisingly small. It was covered mostly in beams that supported the electric LED-display billboards that spanned the sides of the triangular building from the second floor all the way to the roof. A photo that I had been sent less than an hour earlier showed a huge swarm on the end of a massive steel beam. So we knew the bees were there, or had been there just a while earlier. I looked up and half hoped to see them attached to the great New Year's Eve ball—no such luck.

Despite the manager's wishful-thinking protestations that the bees must have left, we hunted around. It took only about a minute for Gus to spot a few bees buzzing around, trying to lead us in the right direction, and Hannah correctly guessed that the swarm had simply shifted position. She gasped aloud when she found the actual swarm. I don't know if she gasped from the surprise of finding it, from the size of the conclave (it was a large swarm), or from the fact that the bees were hanging in space on the edge of a beam with a seventeen-floor straight drop to the sidewalk beneath them. It was truly a stunning site to behold.

Now, one may well ask, why not leave the swarm alone?

Why not let the bees live their lives and go about their business? Both good questions. First, there is the inherent danger, or at least concern, that a swarm in a hyper-crowded locale such as Times Square—through which upward of four hundred thousand people, more than the entire population of Iceland, pass every day—would again relocate and cause mayhem, or, at least, severe inconvenience. The swarm that landed on the hot dog vendor's umbrella down below a year later made a fun story for many, but it did cost the vendor a day's profits. So it is generally better for all concerned that a swarm be captured and safely and appropriately housed.

Risking life and limb for $100 or so's worth of honey bees is something that I cannot easily explain. True, it makes little sense to chance death for something as trivial as a buzzing blur of insects. But when a beekeeper sees a dangerously located swarm right in front of him or her, that beekeeper's DNA is programmed to capture it; normal concerns about mortality simply don't exist, nor do intellectual functions like reason and logic. At least that is how I justify it. Also, I have always been too stubborn to accept that much is impossible. Add to that the fact that, though I knew that this particular swarm was magnificent, majestic, and marvelous, the experience of being there on that roof with the bees seemed to be in its own way a private thing. Meaning, though two hundred feet below I could see thousands of people walking, dozens of buses driving, and scores of taxis zooming, we were far removed from other human contact. No one was looking up. We could easily see Central Park, gaze up and down Broadway for miles, and look into the windows of the buildings opposite. We could hear the faint honking of horns and other nondescript sounds from below. Yet we were alone, isolated with the bees and the wind. It felt like a very personal experience.

So we set about doing the job of rescuing them. From one of the NYCBA apprentices—who was, as kismet would have it, in the area, with his car, and armed with a bee vac—we borrowed a low-suction vacuum specially designed to gently suction up honey bees. Hannah and I strapped on safety harnesses usually worn by the maniacs who regularly changed the lightbulbs on the ads on the building or washed its windows or whatever it is they do. We fastened ourselves to stable points, leaned out, and started the vacuum.

To capture all of the bees, we had to perform mild acrobatics. At one point I needed to sit on the beam and shimmy out over the edge of the ledge, gripping the beam with my thighs. I thought about how those people who used to walk across the wings of planes wore a wire to stabilize them, but how the wire was just a psychological aid because it wouldn't do a thing to help them if they slipped. I decided that when the capture was complete I wanted to take the elevator down as soon as possible—and not the outdoor express one. But there was no incident. For their part, as usual, the bees were passive and uninterested in us, and paid no heed at all until it was too late. They were pulled into a screened cage, where they were safe yet completely flummoxed about what had just happened.

Of course, as we focused on accomplishing our task in a safe and efficient manner, we did not consider that the Reuters offices were right across the street. Hundreds of reporters were working at the same level we were, doing their jobs in a safe and efficient manner, too. Theirs was reporting news. When we were halfway done and I was taking a break on sturdier footing, my cellphone rang and I answered it. It was a photographer from Reuters I had met the week prior at the United Nations, where he was doing a story on the beehives that I managed there. "Would you please raise your right arm?" the voice on the phone asked me. I played along. "Thank you. My colleagues and I are watching you, and they didn't believe me that I knew you."

"So you thought you'd call me when I am trying to hang on to the side of this building covered in bees?" I chided.

"And you decided to answer?" he shot back. Reporters are often smart-asses.

When we reached terra firma again, we proceeded to Bryant Park where we relocated the bees, ignoring the three-mile rule this time. And then we continued working at other apiaries until, mercifully, the sun finally set.

Thanks to our friends at Reuters, our adventure was caught on film and disseminated all over the world. The swarm capture was seen in Korea, Australia, Germany, South Africa, Brazil, and dozens of other countries. It was nice to have our work and appreciation for honey bee rescue recognized on an international level, and it seemed appropriate that the Crossroads of the World was the place to share this moment.

JULY

And your Lord inspired to the bee, "Take for yourself among the mountains, houses, and among the trees and [in] that which they construct. Then eat from all the fruits and follow the ways of your Lord laid down [for you]." There emerges from their bellies a drink, varying in colors, in which there is healing for people.

—Qur'an, 16:68–69

Honey bees, and other bees and wasps, reputedly have a long history of being dragged into warfare by humans and used as weapons. A couple of thousand years ago, the Heptakomites purposely left poisoned honey along the route of their adversaries, the Roman soldiers. This honey, made from rhododendron nectar, was harmless to honey bees but unsafe for human consumption. Hence, when one thousand advancing Romans found the honey, they ate it, fell quite ill, and were easily overpowered by their enemies.* For their part, Romans also are said to have employed bees in their arsenal from time to time, catapulting them into the thick of their enemies. Upon landing, the hives would burst and tens of thousands of angry bees (or hornets or wasps) would engage the enemy on the Romans' behalf. Aside from the Romans, the ancient Greeks are credited to have used bees as tiny soldiers in war. In addition to catapulting beehives over the walls of besieged cities, the defenders of Themiscyra, a Greek town best known at the time for producing exquisite honey, defeated the encroaching Romans in 72 B.C. by funneling swarms of bees into the mines beneath the walls of the city—straight into the faces of the enemy. Fast-forwarding another thousand years or so, Richard I of England is alleged to have used bees as catapult-launched bombs against the Saracens during the Third Crusade in the twelfth century. There are examples from the American Civil War of bees and apiaries disrupting battles and troop movements; they were used, at least on one occasion, as a practical joke against the captain of a regiment. "I recall an incident occurring in the Tenth Vermont Regiment—once brigaded with my company—when some of the foragers, who had been out on a tramp, brought a hive of bees into camp, after the men had wrapped themselves in their blankets, and, by way

---

* "They gathered up great numbers of wild honeycombs dripping with toxic honey and placed them all along Pompey's route. The Roman soldiers stopped to enjoy the sweets and immediately lost their senses. Reeling and babbling, the men collapsed with vomiting and diarrhea, and lay on the ground unable to move. The Heptakomites easily wiped out about one thousand of Pompey's men." —Adrienne Mayor, *Greek Fire, Poison Arrows, and Scorpion Bombs* (Woodstock, New York: Overlook, 2003).

of a joke, set it down stealthily on the stomach of the captain of one of the companies, making business quite lively in that neighborhood shortly afterwards."* But it wasn't all fun and games, of course. A Union soldier named Lieutenant Robertson from New York's Ninety-third Regiment, while in Virginia, recounted, "To advance was impossible, to retreat was death, for in the great struggle that raged there, there were few merely wounded. . . . The bullets sang like swarming bees, and their sting was death."

During the Great War, the Germans gained an advantage during the Battle of Tanga, also known as the Battle of the Bees, fought near the slopes of Mount Kilimanjaro in what is now Tanzania, when several beehives were disturbed by gunfire and the bees angrily swarmed the area, particularly attacking the British Ninety-eighth Infantry Regiment and the Loyal North Lancashire Regiment. In his memoirs, the German commander, Paul von Lettow-Vorbeck, described the accidental episode as "decisive" for his victory. In Southeast Asia, the Viet Cong used stinging insects to fight, using nests as projectiles and in traps to be stepped on. Honey is sweet, but the bee stings.

In July 2005, I was working with bees far away from Manhattan rooftops and Brooklyn community gardens. Under the auspices of a State Department initiative and embedded with the U.S. Army, I was teaching beekeeping in Iraq.

While Iraq during that time was as brutal as most people

---

* Billings 1887, from *Historical Natural History: Insects and the Civil War* by Gary L. Miller

might imagine, it was also filled with good people, life, joy, hope—and to some extent, honey bees, and a need to bolster the honey bee industry for the purpose of helping the Iraqi people feed themselves through increased pollination via more and healthier honey bees. So I was all in.

I moved around a lot, from Baghdad to Erbil, Kirkuk, Sulaymaniyah, Mosul, and several other places. Some locations were rough and tumble, and in some, one would hardly know there was a war going on. One relatively peaceful region was Dohuk, located in the northernmost part of the quasi-independent Kurdistan, in northern Iraq. It lays claim to beautiful mountains, hills, and lakes, and also produces some of the most delicious honey—yellow star thistle blended with eucalyptus—that I have ever tasted in my life.

To get to Dohuk from Erbil, our convoy drove at breakneck speed, mostly because we were passing through the outskirts of Mosul, a city of infamous reputation and one that had been much in the news at the time as a stronghold of the opposition. The road was bumpy and curvy, and by the time we arrived, I felt that I would vomit my organs onto the front of my body armor. Motion sickness is my kryptonite.

After three hours of being tossed around, we finally arrived in Dohuk. We had a scheduled meeting at a bee farm, but I asked to make a quick stop first to compose myself; a cup of tea and a bite of bread, I thought, would do wonders for my state. We entered the parking area of a strip mall of sorts, and my ever-present security detail of about a dozen men, some American and some Kurdish, and I entered a café. First they told us there was tea, then they told us there wasn't, then we were informed that there was but we would have to wait fifteen minutes for it. We decided to use the time to explore the grocery store next door.

During my time in Iraq I was astonished by how much attention was given to my movement by the security team, and so I tried not to deviate from the schedule, as it threw them into high gear. They would stop traffic whenever we entered or exited a building. If I had to use the restroom, they would surround me as I approached it, and then clear it out and maintain watch outside so that no one could enter while I was performing my ablutions. When we entered the only super-market in northern Iraq, located in the strip mall with the café, it was cleared out except for staff so that I could shop. I bought some Pringles-like chips.

I imagined being back in the States and going to a Stop & Shop, CVS, or Walmart, and having guards block the parking lot, clear out the building, watch all the exits, and surround me as I bought potato chips. That, from time to time, was my life in Iraq. It was a strange blend of high tension and complete relaxation. And while the security measures often felt unnec-essary and cumbersome, they did allow us to avoid delays due to long lines. That was my positive spin on the situation.

Refreshed in all ways, we drove the remaining distance to the bee farm. To get there, we crossed a massive dam formerly known as Saddam's Dam and now once again called Dohuk Dam. The stone dam, adorned with a large painted-on Kurd-ish flag, is home to one of the most gorgeous vistas one could hope for. It reigns over a massive blue lake cradled between impressive hills and mountains. The view is a sight to behold.

A group of about twenty beekeepers awaited our arrival. They were sitting patiently beneath a canopy of natural materials, sipping sugary tea from tiny glasses on tiny saucers. When I arrived, the head beekeeper took me by the hand and led me to the others. I shook hands with every one of them. Some clasped my hand between their two hands, released their grip, and then touched their hearts. Many also made attempts to introduce themselves in English. I tried using my extremely limited Kurdish during my greetings, which was oftentimes met with a look of confusion, but also with smiles. We had an interpreter, but he wasn't necessarily well trained. We worked with what we had. Everyone had a good attitude and we were gathered for a common purpose—to promote healthy honey bees.

When it comes to beekeeping, while I have the spirit, the passion, and the technical know-how, the man who taught me everything I know about bees, my father, holds more information about bees in one hair of his white mustache than I will ever know. I also sit on the shoulders of great beekeepers like Dr. Thomas Seeley, Dr. Larry Connor, and Dr. Dewey Caron, people who have taught me along the way through their books, talks, conversations, and friendships. So, bearing that in mind, imagine how futile I sometimes feel when I have a group of hardworking, impoverished beekeepers sitting in front of me, hoping that I hold the magic key to their varroa mite problem, their wax moth enigma, or their difficulty in raising a decent honey harvest.

Especially varroa mites! Those little buggers had killed about a third of the colonies in North America the year prior to my visit to Iraq, and if I knew the secret to destroying them, be sure I would share it. The parasite *Varroa destructor* attaches itself to the exoskeleton of the honey bee and hangs on, sucking and feeding on the fat found in the abdomen of the bee; if a human being were to have equally proportioned creatures dangling and feeding on the liver, it would look as if big red basketballs were affixed to him or her. Varroa came to the States around 1986 and has radically changed beekeeping here and wherever they rear their nasty little red heads. When I say "changed," I mean that at this point, about 30 percent of American honey bee colonies are decimated per year, and the number one cause of their demise is, in the opinion of the great Dr. Dewey Caron, varroa.

My beekeeping knowledge and experience, as limited as it was, could at least help these beekeepers increase their yield of sweet liquid gold and stave off the more devastating effects of varroa. I wouldn't have made the trip if I'd thought I couldn't. But there's no magic or simple solution. And one must possess an open mind and be willing to accept that, perhaps, there are practices that need to be adjusted. Not that my practices are flawless, but I do have the benefit of further-reaching experience, and experts with whom I am in contact.

Among the difficulties I run across, especially when I travel, are folkloric ideas that the rural beekeeper simply will not relinquish. An example: The beekeepers in front of me told me that there was a general feeling among those in their village that bees can recognize the scent of their keeper. There are many beekeepers in the United States, too, who believe that their bees "know" them from their smell, their touch, and their manner—similar, I guess, to the way a dog recognizes its owner. Bees do have an extraordinary sense of smell, after all.

Beekeepers tell me all the time that they do not wear veils when doing routine inspections, and that their bees never sting them because they know them and they have an unspoken understanding. It is a notion as charming as it is false.

The moment the notion of bees knowing the scent of their keepers goes too far is when, for the purpose of assuring that one's bees are indeed familiar with one's scent, one places a pair of used undergarments into the hive. Besides the obvious, I have a couple of problems with this practice. First, I imagine it is a daunting task to fill hundreds of hives with underwear, for just how long might it take to amass so many pairs of used skivvies? And second, I cannot imagine that the underclothes exert any sort of positive influence over the honey. On the contrary, I think the noisome addition to the hive would distract the girls from their tasks. What decent woman likes to work with a pair of men's dirty drawers in her face?

Another interesting discrepancy between the world as I know it and the world as the beekeepers in this region knew it involved the behavior of the queen. I was repeatedly and adamantly informed that I was wrong about the queen's mating flight. To review: The queen bee mates once in her life, on what is called a mating flight, and on that flight she mates in the neighborhood of fifteen times. That is, she copulates with that many drones, whose barbed penises are then torn from their bodies, ripping out their entrails. Then the drones fall to the ground and die. After this busy afternoon, the queen stores the sperm of her suitors for three to five years. Most queens can lay up to two thousand eggs a day when necessary for two years, with numbers dropping off somewhat thereafter.

I don't offer this information as opinion or conjecture. These are the facts.

The very idea that the queen, as perfect as she is, would have relations with so many males was unbelievable to some of my Iraqi friends. "Is she a *whore?*" one particularly impassioned man demanded, nearly screaming. He enthusiastically refused to believe a pristine female—such as his mother or his favorite wife or his queen bee—would keep company with so many men. So there is that element as well to contend with: cultural differences. There are a lot of intercultural communication issues that come up in my Bees Without Borders work, and I am sure that many of the people with whom I deal overseas are as baffled by my everyday comportment as I often am by theirs. Regardless, my underwear stays out of my beehives.

We talked more about bees. Someone brought me the corpse of the dreaded European bee-eater, *Merops apiaster.* Even in its poor condition, it was easy to see that it had once been a handsome specimen. *Merops apiaster* is a beautiful bird, richly colored and slender, with a black beak, green wings, and a combination of browns, yellows, and even gold in its feathers. While beautiful, it is thought to not only eat a great many honey bees—and indeed, it does eat them—but it is also believed by many beekeepers that this feathered fellow will, with his mere presence, intimidate the honey bees into staying inside lest they become a repast for these migrating beauties. The truth is that even in areas where these birds nest, fewer than 1 percent of the local bee population ends up as lunch for the bee-eater. Still, try telling that to a group of Kurdish beekeepers or any rural beekeeper from West Asia all the way to Africa. Most beekeepers in the region would seem to disagree.

Interestingly enough, there is at least one place on earth where honey bees are purposely cultivated as a meal for bee-eating birds. Founded in 1899, the Bronx Zoo currently boasts nearly two and a half million visitors per year. There are two beehives kept atop the World of Birds at the zoo—two colonies of honey bees that have been populated on and off for the past dozen or so years, there primarily to provide sustenance to a specific group of birds on the ground floor of the building. The white-throated and white-fronted bee-eaters (*Merops albicollis* and *Merops bullockoides*, respectively) feed in part on those honey bees raised on their roof. Two beehives, in traditional Langstroth beehives that are equipped with small pipes as entrances, are on occasion fitted with tubes to those entrances that lead into a small trap. Once a couple of hundred bees fly out of the hive and into the box, the box is sealed and brought downstairs. Then that box is taken into the small arena where the hungry birds reside. There it is opened, and the unlucky honey bees become an afternoon repast for the hungry birds. The public is invited to watch the action.*

---

* As if that isn't enough, from time to time the zookeepers take slabs of honeycomb and leave them scattered in the bear pen as snacks for the brown bears. Bears love honey—and honey bees and larva. So the Bronx Zoo bees, directly and indirectly, are filling the bellies of the birds and the bears. But there is more! The best part is the manicure services offered to the bears. Trimming a bear's nails (claws?) can be a challenge. The folks at the Bronx Zoo fill an oversized baby bottle with honey from which the bear will suckle and slurp out the sweet thick liquid. That bottle is held high up behind a fence, enticing the bear to stand up and place its mitts on the barrier. While said bear is feasting slowly and greedily on the sugary beverage, its paws rest on the fence, nails through the openings, which are trimmed without any trouble at all. If anyone knows of a more adorable way to trim a bear's claws, please send a note to my publisher and let me know, too.

But back in Iraq, we went on to speculate on the cause of the death of some local beehives, discussing the application of a certain chemical that some foreign company has distributed liberally in Iraq, but which has been banned in the States and many other countries with even minimal regulations. They asked me why, if it had been banned elsewhere, was it given to them? I answered as best I could that the reality was that some company had manufactured it and wanted to profit from it, and when they couldn't sell it in the United States or another reasonably well-regulated country, they'd probably sought out a market where there were fewer regulations and safeguards. Iraq is not in a position at the moment to dedicate its resources to checking the safety of every medicine, and so these chemicals get dumped here and elsewhere. I urged the beekeepers I met in Kurdistan not to use certain products I knew to be harmful, though I understood that it would be hard for them to resist, since they need to battle the varroa mite problem, for example, somehow. In short, over time, these struggling beekeepers will poison their bees and strengthen the varroa mite by using and misusing the toxic chemical.

One man suggested that burning lavender inside the hive would kill adult varroa mites, and that by using that method every few days, one would kill all of the varroa mites as they came out of the cells. I have also heard from a beekeeper in Japan that a lavender patty of some sort was being devised. I find it interesting that two similar folk remedies have evolved, and I hope that one of them will prove effective. Unfortunately, the evidence weighs against it. The latest natural remedy for varroa mites that I've heard involves mushrooms. Time will tell.

During a break in the meeting, the head beekeeper,

Khorsheed Ahmed, led me around his apiary. One thing I noticed straight off: The poor yield of the hives must have had a lot to do with the fact that they put far too many colonies in an area that cannot support that many foraging bees. When there are simply too many bees in one place, the nectar sources will run out, and the bees will have weak hives and poor honey yields. Also, there was an inadequate nearby fresh water supply. The former problem could be solved. It is important to bring water to the hives, especially in dry rural areas.

Eventually we moved from the hillside to a nearby restaurant, where my organization bought lunch for the forty people or so in our group. We ate traditional Kurdish food and continued chatting about bees. I noticed a young beekeeper, about sixteen, which was a rarity in what is usually an old man's vocation in this country. It turned out that he was there with his father, who has been teaching him about bees since he was a much younger boy. This reminded me that many—indeed most—of the beekeepers had learned their craft from their own fathers or uncles. Of course, this put me in mind of my own father and his tutelage, which has brought me to wherever I have gotten in the bee world and beyond. I tried to explain that to my new friends, but I had to abruptly stop my talk about it, for despite myself I realized that I was welling up. Surprised by this swell of emotion, I once again realized how grateful I am to my father for teaching me about bees, and for being a great father. A sentiment and bond that was clearly understood and shared by many present that day.

Eventually we retired to a secure hotel, where we met some more beekeepers from a different group. We all spoke and sampled different honeys, and then they were on their way. I had an appointment scheduled with the minister of agriculture the next morning that I wanted to be ready for. Not that I had any clue what we would be meeting about. My primary bodyguard, a man named Dave from Staten Island, and I decided that since the hotel was heavily guarded we would send my personal guards out to enjoy their evening, and he and I would stay in the hotel and have dinner. With twenty dollars from me, which was more than enough for several beers each for the guards, they wandered off, and Dave and I sat in the outdoor restaurant watching a huge moon rise, sipping beers, and tossing scraps of our food to the scrawny cats that meandered around the tables.

While we were eating, a fellow named Marwan came by. Marwan worked with my organization, and he let me know that he'd had no luck finding an interpreter for my meeting with the minister of agriculture the next day. Without one there was little I could do, so I had the idea to hire one of the receptionists at the hotel. I found a woman who had worked as an interpreter for some news organizations in the past, and though her English was not stellar, it far surpassed my Kurdish, which consisted of about ten words. Problem solved, I thought. I went to my hotel room and watched BBC to learn more about the war that was raging elsewhere in Iraq.

Not long after I'd gotten back to my room, Dave radioed to tell me that he was around the corner sitting with a group of three tourists, members of a little society that promotes and enjoys traveling to war zones and other dangerous places. It all started with a book (one that I happen to have read) called *The World's Most Dangerous Places*. The book's most avid fans evolved into a cult of people who like to travel to dangerous areas and write about that experience. But as far as Iraq is concerned, visiting Dohuk and then leaving the country is perhaps the equivalent of going into a house of ill repute to use the bathroom, and then quickly departing without partaking in any of the other offerings. You don't really see any action in either case. Still, these three, two Brits and an American, were very chummy and we shared beers for a few hours.

"I work for the government," confided Charlotte, a Rubenesque middle-aged bleached blonde whom we had seen earlier in Dohuk, either leading or being led by her big red suitcase on wheels. It turned out that she worked for a county clerk's office somewhere in the Midwest. A pleasant woman, she was excited to be in Iraq for the day. The two fellows, Ian and John, spent a lot of time in Turkey, I gathered, but other than the fact that one of them worked with computers, I did not catch much about their line of work. They didn't believe that I was in Iraq teaching beekeeping skills.

"That is the worst cover story I have ever heard!" insisted Ian.

"So bad it must be true," I countered.

They were not convinced. "So why all the firepower with you?" John pressed.

"In case a bee tries to sting him," Dave answered. More drinks and laughs ensued. For those few hours we could have been just about anywhere.

The next morning I met the interpreter, Berivan, downstairs, and we all drove off in the convoy to the Ministry of Agriculture. It was only about five minutes away, and during that time I learned that in Kurdish, Berivan means "girl who takes milk from cows or sheep." I was about to learn that milk-maids are not necessarily the best interpreters.

Also, it was the case on this day, as on many days over the months I was in Iraq, that there was no clear plan or reason for any of these meetings. They just seemed to enjoy having meetings. Sometimes I knew why I was walking into one, and sometimes I didn't. Often the meeting was about what we would discuss at the next meeting.

We were ushered into the minister of agriculture's office and took our seats as he was talking on the phone. Where we sat there were two air conditioners that blasted out nothing but hot air—much like the minister himself, as it turned out. A family of birds was clearly roosting in one of the AC units. I asked my interpreter if he was on an important call. She listened, and then reported that he was talking about shoes, at which point I noticed him regarding his shoes more than one normally might. Then, after a full twenty-five minutes of me listening to a man talk about footwear in a language that I could not understand, Mohammed Sherif gave us his full attention.

That lasted about one minute until another phone rang, this time his mobile. He picked it up and started talking again, obviously delighted by something or someone. Mercifully he was on this call for only about five minutes, and then he imme-diately launched into a passionate speech about how Kurdish honey is world famous, and how it is the purest and best in the world, and how the Qur'an tells us that honey is a medicinal food, and how there are two kinds of honey, natural and chemical, *blah blah blah*.

I was bored out of my mind as he spewed his rehearsed

rhetoric, but I nodded and assumed a serious expression and did my best to be polite. I could hear Berivan's interpreting just fine up until it was important that I hear her, and then suddenly she decided to be coy and whisper. This had also happened with our translator the preceding day, and I wondered if the technique was somehow part of their training. Between my interpreter's sudden and extreme drop in vocal volume and the irritating marketing pitch that the minister seemed to be making, my eyes were glazing over. I was wondering what I was doing there when the minister surprised me by barking in English "QUALITY OF HONEY!"

Around this point his phone rang again and he slumped into his chair to answer it. Call number three seemed like a genuinely grave matter. After a minute or two, my interpreter told me that it sounded as if he was doing some home remodeling and something was going wrong.

Not long after he got off the phone, it was time for us to say goodbye and head to the main purpose of our visit to the ministry—a meeting with inspectors for the region, under whose purview fell beekeeping—though, like their New York City Department of Health counterparts, not one of the inspectors was a beekeeper.

At this point, right when the true work of the day was about to commence, my interpreter's phone rang. She answered it and then quickly scribbled on my notepad as she listened to the voice on the other end:

"Sorry, phone come to me argent I most be I go, I am apologize If you don't mind."

Translation: *Andrew, you are about to go into the big meeting, and you have no interpreter. Good luck! You're screwed!*

So Berivan the interpreter was off. Soon thereafter I came to wish I had left the building with her. These bee inspectors were not there to have an amicable gathering. They just wanted to demand money and to complain. I first knew something was wrong when they did not offer me the best chair in the room. Everywhere I had been in Iraq, I had been treated very well and with respect, as it was customary to treat a guest in that country. Especially a guest who holds the key to a lockbox of cash. Mind, I don't need the finest chair, or always accept it. But a guest is generally offered it regardless. Not this time. Instead I was offered a rickety plastic chair next to the cushy one where I had placed my bag, which someone had moved onto the floor. The nice chair was now filled with the posterior of a government worker, who appeared unmotivated to be elsewhere. No one offered me a drink. I had never been to a meeting in that country where I had not been offered refreshments. Instead, through the impromptu interpreter someone rustled up, I started to hear the first of a dozen demands from this surly group of government workers.

These inspectors had been living in something of a welfare state for quite some time. I didn't mind hearing suggestions or pleas for assistance, since that is part of why I was there. However, the things this group of petty government officials were demanding included a factory to sew overalls for beekeepers, a factory to refine wax, a factory to make hives and hive tools, a university and a laboratory to test the pureness and quality of honey; quarantine stations at every border in the country to prevent pests from coming into the country on the backs of bees; a fleet of ships and planes to transport and export their honey—no worries to them that the area is landlocked; I am sure they meant to ask for a canal.

I grew weary of it and their attitude early on. One man seemed to have been elected spokesperson, and he was shouting these demands for the entire world to hear. My woefully inadequate interpreter could hardly make out how to tell me anything. At one point I asked them if they would like a ladder built to the moon. This was interpreted (maybe—something was interpreted) and they seemed to discuss it, and finally the response came that there was no need for one at this time. I told them that they had about as much luck in getting that ladder as they did in having these outrageous demands met. I asked them what they had done in their professional capacities to better the plight of the beekeepers in their country. Precious little, I answered for them. We give grants to those who demonstrate initiative, I told them, and who make reasonable proposals. They were talking about hundreds of millions of dollars. One man shouted, "What do I do when I have five tons of honey? How will you help me export it?" I responded, "When you have even half a ton of honey to worry about, we will talk about it."

I had hoped to discuss local projects with area beekeepers, teach them how to manage diseases, and perhaps set up some training programs and shared equipment facilities—something useful to the actual beekeepers, not just to the pockets of the pencil pushers in charge. Not another well-funded program for them to fleece.

I decided to leave fifteen or so minutes into the melee. There was no point, as I saw it, to this meeting. These men were not beekeepers; they were petty bureaucrats looking for something for nothing. Middlemen scheming to personally profit under the guise of government service. As I started to leave, one man was apparently not finished shouting. This was for the sake of his colleagues, no doubt, since no one was interpreting any longer. He moved just a little too quickly toward me and immediately became better acquainted with one of my armed guards. I smiled and walked out, wishing I always had these fellows with me. Especially knowing how opening my mouth sometimes got me in trouble.

We returned to the hotel to eat lunch and check out. These things accomplished, everyone piled into the vehicles to drive up to the dam once again. Once there, we drove a few more minutes to a picturesque spot near the pristine blue waters. We all got out of the vehicles, walked a few hundred yards down past the DANGER! NO SWIMMING! signs in Kurdish, and disrobed and jumped in. Since we could not technically read the signs ourselves, Dave and I reasoned, we were in no danger. The weather was hot, well over 115 degrees Fahrenheit, and the water was cool and clear and perfect. It was a nice diversion, and all of the guys seemed to have a good time. In thirty minutes we were back on the road and looking to get gas before heading back to Erbil. Dave, who had been wounded in a car bomb attack about six months earlier, said, "This has been my best day in Iraq so far."

I probably need not mention that things usually start to go wrong after such a remark.

We arrived at the gas station. For reasons unclear, the owner did not want to sell us gas. It was not because we had jumped the half-mile-long line—all military and paramilitary vehicles do that. I never learned the exact reason why, but he did not want to sell to us, and he said so plainly. Dave was angry. We needed gas, and we needed to get on the road to get past Mosul and get home before dark. Dave threatened to shut the gas station down, to set up his vehicles at either entrance and make sure the owner got no business whatsoever. The clock was ticking, and still the owner would not budge. Tempers flared and the driver of my vehicle got out and started shouting at someone. We had a mix of Kurds and Arabs and Americans, and no one seemed to fully trust, and certainly did not like, the others. It was becoming very tense, but ultimately we were served, we paid, and we left.

Tensions continued to run high, however, and Dave reminded the drivers to go easy on the drive home. Then, naturally, two of our vehicles collided right outside of Mosul— one apparently had been following the other too closely. We eventually made it back to base, but it was not a great ending to what was mostly a long, unproductive day.

We arrived at our compound just as it was getting dark. We were muddy beneath our clothes, as the lake had been sur- rounded by deep, soft mud that squished between our toes. We were all hot and tired. I was also hungry, so I went to the kitchen, where I found the cook cleaning up. She was a woman from Baghdad, an Arab, not a Kurd or Assyrian like most of the people I had encountered in Erbil. I had a hankering for couscous. I'd had couscous with every meal in Dohuk, and I had seen some there in the kitchen before. But when I asked for it, she looked taken aback. I kept repeating "couscous," and saying how good it was and how much I wanted it. I may even have rubbed my stomach and made an "mmm, mmm" noise, which to my mind was an international sound and gesture for "Good eatin'!" She left, upset.

A short while later I discovered that outside of the United States and the Maghreb in North Africa, "couscous," or at least, "cous," in Arabic refers to the female genitalia.

No wonder I didn't get any, asking the way I did. I showered, went to my quarters, wrote my report, and went to sleep.

I'm always in favor of helping others through the vehicle of the miraculous honey bee, particularly in difficult areas during hard times. It was a tough trip, though. I discovered no healing waters in Iraq, in spite of the perfection of our swim behind the Dohuk Dam.

The time I spent in Iraq was mercifully conflict free for me, compared to the experiences of many others during the same general time frame. I was able to meet and hang out with some excellent people. My friend Marwan, thank G-d, avoided some of the worst atrocities committed against the Yazidi and is now living in Sweden, with family just one town over. Dave now lives in Wilkes-Barre, Pennsylvania, and has added "grandfather" to his list of job titles. One of my former Assyrian interpreters and her family are in Chicago and have recently welcomed a new baby to their family. One of the Iraqi beekeepers who was also a dermatologist went on to take over the century-old apiaries of the late Brother Adam at Buckfast Abbey in Devon, England. And so, for the fortunate, life continues, in ways and directions that none of us could have expected.

* * *

Aside from being used as tools of war, honey bees have been used for healing the wounds of combat as well. In fact, anyone who suffers from a stressful job or situation may benefit from working with honey bees. The experience of opening the hive, smelling the heavenly scent of the freshly made honey stuffed into immaculate beeswax, and hearing the steady hum of thousands of workers striving toward a common goal is soothing and therapeutic. Since the return of soldiers from the Western Front following the Great War, beekeeping has helped people find a semblance of peace and continuity after horrific experiences.

The Federal Board for Vocational Education, whose sole mission was to assist disabled soldiers, sailors, and marines in finding work, put out an "Opportunity Monograph" in 1919, the thirty-seventh in a series, this one specifically to promote beekeeping for injured and maimed servicemen. It urged them to "make beekeeping [their] life work," enticing them with "Uncle Sam foots the bill" for most or all expenses associated with their training. The text reads, in part, "Training counts. You know it counts, for it was training that helped you beat the Hun."

The driving force for this particular program was to help physically disabled veterans to gain a means of income generation. The monograph illustrates two World War I veterans—Mr. Nicholls, who lost both his legs, and Mr. Donnegan, who came home missing one arm—who returned to the United States and in time took up beekeeping as their vocations. "It is especially attractive . . . to those who have . . . lost one or more limbs. . . . A beekeeper should, however, have one good hand and arm." (It noted, in the margins, "If you need a new arm or leg, that will be provided, one for Sundays and one for the workshop. You can play the game with it as well as with the one you left over there, and it won't hurt when you pound your thumb or get it broken . . . and it is warranted against rheumatism.")

Though this pamphlet did not address the mental wounds and anguish these servicemen experienced, there must have been something to having honey bees act as agents of positive change, because when World War II ended, there was another Veterans Administration–approved program to assist returning disabled veterans to learn beekeeping. This was more of a general farming program, aimed at those men who had been blinded. The headline of a 1947 article in *The Baltimore Sun* reads THEY MILK INVISIBLE COWS. At Barnes Farm School for the Blind in Henniker, New Hampshire, the veterans were "taught enough planting, animal husbandry, beekeeping, poultry handling and other farm tasks to qualify as competent farm workers." A blinded former GI and city boy named Joe Lysak spent two years learning how to be a farmer, rather than live off the total disability payments he could easily have taken from the government. He concluded that his example "should knock into a cocked hat that sarcastic old saying about 'the blind leading the blind.'" And the notion of a blind person tending honey bee colonies is not new or far-fetched. François Huber (1750–1831) was a Swiss entomologist who lost his vision as a teenager. He is renowned for his breakthrough work with honey bees, all of which was entirely based on observations spoken to him through his wife, servant, and many assistants.

Today there are still programs that use bees and

beekeeping to help those with post-traumatic stress disorder find an equilibrium that has evaded them, including Michigan State University's "Heroes to Hives" program. I could not put it better than former army private first class Adam Ingrao, who said, "When you are working beehives, if you are not thinking about what you are doing—if you are not in that space—you tend to get stung." Working with bees is a way to leave baggage behind.

Personally I find it virtually impossible to think of anything other than the honey bees in front of me when I am in the midst of a hive inspection. The organized chaos that is a constant in the hive is wondrous and eliminates all other concerns for that spell of time one is among them. I find working with the bees meditative, reflective, and healing.

AUGUST

One can no more approach people without love than one can approach bees without care. Such is the quality of bees.

—LEO TOLSTOY

All through the swarm season of 2012, and even the year before, I received calls about swarms in Corona, a neighborhood in Queens. There seemed to be too many for one area without a central source, but I couldn't find the root of the problem, and the list of local hives provided by a sympathetic source inside the Department of Health did not reflect any sizable apiaries in the vicinity. It was vexing, but there were plenty of other bee problems and mysteries to solve, honey to harvest and hawk, and rooftops apiaries to cultivate.

August is what is known as the dearth period. The dearth

is when nectar flow for the bees is at a minimum. During the dearth, there are few resources from which the honey bees may draw their food, and so they begin to change their operations a bit. As the year progresses, the queen continues to lay fewer eggs, so that there are fewer mouths to feed the smaller quantity of foodstuff available. The guard detail at the front entrance to the hive is beefed up to deflect the increased number of robbers who might want to make a grab for winter surplus. And as implausible as it may seem to sweating humans, it is during the dearth, when temperatures are at their highest, that bees begin to batten down the hatches and prepare for the long, hard winter. The dearth is also when the bees become most susceptible to pests such as varroa mites, the parasite that leeches on to bees when they are still in their cells in their pupal stage.

In August 2012, the dearth period led to one of the strangest, messiest urban beekeeping events I've ever encountered in all the years that I've been keeping bees. It's a tale of excess, neglect, bee abuse, bait and switch, and karma in action. It's a story that's been told far and wide; the scene even earned a mention in *Ripley's Believe It or Not!*, which would have made me unspeakably proud in my adolescence, and still made me smile decades later.

On July 23, I received word from a man in Queens named Dr. Dalton Garis regarding some beehives that a client of his realtor wife was keeping at his home in Corona. Garis, then sixty-five years old, had earned his PhD in food and resource economics from the University of Florida. He'd found the New York City Beekeepers Association online and sent this email:

A friend of ours has extensive beehives now producing honey. He must sell his house and needs to find a new home for the bees. The hives are behind his house. . . . Call me anytime at these numbers below if you have any suggestions.

I phoned Dr. Garis, and he put on his wife, a member of the ever-expanding Chinese community in that area of Queens. What she lacked in English proficiency, she made up for in volume and speed. She also liked to repeat herself and did not consider it important for a conversation to be a two-way endeavor. It was a painful introduction to the Corona bee project and one I should have seen as a harbinger for what was to come. Sadly I did not.

Mrs. Garis said, or at least I understood her to say, that she was a realtor and that one of her clients had "four or five" beehives in his backyard, and a lot of honey to sell. This honey part matched what her husband had written via his email. Prior to my call, I wrote to Dr. Garis and let him know that I could not help with the honey (for various reasons, mostly the fact that I had enough honey from my own beehives), but that I could probably assist with relocating the beehives. Dr. Garis replied: "He must sell everything. Price is negotiable. My wife can act as translator."

To which I replied: "I am only trying to help—I do not need the bees or the honey—but I can probably find someone. I assume this needs to be done immediately, yes?"

It did. We agreed on a day and time for me to head to Corona and scope out the situation. A few days later, I arrived at the site with my friend Tom Wilk, who was also a Queens resident, an enthusiastic curler, and a budding beekeeper. Tom, a native of Greenpoint, Brooklyn, had been keeping bees for only about three months, but he had already shown great heart and a strong work ethic. He had helped me move live hives, harvest honey, and capture swarms, and had been the recipient of more stings than I could count, taking them all with only minimal and understandable complaint. As a wholesale wine rep for more than twenty years, Tom also keeps me saturated in fermented delights, so he is a super good fellow to have at the arm.

I parked my pickup, and Tom and I walked toward the address. We could not have been prepared for what we saw. In this tightly populated neighborhood, on a narrow strip of drive-way leading to a two-story brick house, was a rickety mound of active beehives spanning the entire width of the pavement. There were not four or five beehives; there were more like forty-five; I had apparently misunderstood Mrs. Garis. The beehives were in shambles. They were stacked atop one another in tee-tering piles as many as four colonies high with no fresh water source. Since the boxes faced all four directions rather than one consistent one, many of them faced the neighbors' doors and windows. The house to one side was a multiple-family residence with doors for different apartments in the front, the side, and the rear of the building. Some people would have had to walk at least forty feet through that horrid scene every time they entered or exited their home during daylight hours. It was a honey-making sweatshop. Even in poverty-stricken places like Nigeria, Cuba, or the mountains of West Virginia, I had never seen an apiary as disorganized as this one. The post-earthquake apiaries in Port-au-Prince were in better shape.

As it was a warm day, there seemed to be bees exploding in every direction out of the hives, but unlike a real explosion, the momentum of this bursting forth never slowed. The activity was all within a few feet of the public sidewalk and just two steps from neighbors' windows and walkways. It turned out that multiple complaints had been made to various city authorities, but no one knew what to do about the situation and the conglomerate of complaints never got to the right place. Weeks later at the request of a pleading next-door neighbor who spoke only Spanish, I personally called the local precinct, was treated poorly, placed on hold, and then disconnected. Not that the local precinct would or could have done much.

The realtor, along with the owner of the house, Yi Gin Chen, then fifty-eight, came out and we all talked about the bees and the honey. "Hello, Mr. Chen," I said and smiled. "You just call him Chen," said Mrs. Garis. He took us inside and showed us that he had about thirty pails of honey containing about sixty pounds each. We were given samples of the honey. It tasted exquisite. It had a strong minty essence and was thick and complex. The bees had most probably availed themselves of the nearby Flushing Meadows Corona Park, the fourth largest park in the five boroughs, for the lion's share of their nectar. It was gorgeous stuff.

I did not need any of the honey, but I called a friend of mine who runs a family bakery in Queens and told him how great it was. He was interested in reducing his carbon footprint in terms of sweetener, and he loved the idea of using honey from Queens. We struck a deal with Chen. This was a laborious process—there is a reason why, at my honey stand, I have a sign that reads in English, Chinese, and Russian, NO BARGAINING. Those second two language choices are not random but are directed at cultures where bargaining is inherent and expected. But I cannot afford to turn every simple transaction into a long-winded exchange. Plus I'm not as bargaining-savvy as some and I would likely lose every transaction.

My friend ended up buying twenty buckets of Chen's product. It seemed like a great arrangement in that Chen got to unload some of his honey, and the bakery got a source of local sweetener at a reasonable price. So money was counted out and handed over, and buckets were loaded into the back of my truck. I did not personally load the buckets; Chen and his assistant insisted on doing that. The honey was dropped off at the bakery, where their staff unloaded it. And all parties were smiles and handshakes.

Next was the problem of the beehives. Chen wanted to sell the beehives. "We understand," we told him. "We'll help," we told him. "How much would you like to sell them for?" This is when things got more difficult. He did not want to say what he wanted to sell them for. He did not want to give out his telephone number or address to anyone, or for me to share it. He wanted to be offered a price for them and negotiate from there, but he wanted all arrangements to go through me or the NYCBA. I explained that the best I could do would be to post an ad on our social media accounts to alert people to the hives for sale. I warned him that finding an urban buyer who had the knowledge and means to safely secure and move about fifty beehives would be nearly impossible.

My suspicions had been rising since the beginning. When I asked a simple question like "How long have these beehives been here?" Mrs. Garis and the man would have an exchange that lasted much longer than a translation of my simple question warranted. This happened often. Chen was pleading for help yet seemed wholly unwilling to allow himself to be helped. For the sake of the bees and safe urban beekeeping I wanted to assist; I could not ignore this calamity of an apiary, the welfare of the bees, and the concerns of Chen's neighbors. But there was only so much that I could do.

Then I got a text from my friend at the bakery. He was displeased. We headed back over there.

Even the delicious smell of freshly baked bread could not assuage the bitterness that sat in wait at the bakery. It seemed that the so-called honey that Chen had sold was nothing like the mouthwatering substance we had sampled. It was extremely thin, sickly sweet, and tasted much more like syrup than anything else. It had a faint whisper of the essence of the original honey in it, but tasted and appeared as if it had been diluted down to something like one part honey and four parts simple syrup. It was an infuriating situation on several levels. First, the gorgeous honey was ruined. Next, even though Chen later denied messing around with the honey, it was clear to me that he was a fraud. He had cheated a local small business owner and made us unwitting accomplices in the swindle. Finally, he had wasted many hours of our time and misled us as well. When we returned and confronted Chen, he played dumb and told us to go away.

So we did. The baker lost out and I'd wasted half a day, but I essentially let it go. Still, the ad that I had placed yielded some results, as a fellow responded that he wanted to purchase the hives. I passed this on to Dr. Garis. A price was arranged, and lo and behold, Dr. Garis contacted me again and asked me if I would prepare the hives to be moved—screen them in, crank-strap them in preparation for their move. Despite my mistrust of Chen, I agreed for two reasons: One, the welfare of the bees was at stake. If this were to be done incorrectly, the bees could die, and/or there could be a massive exodus of bees, which would be bad on multiple levels. Two, Dr. Garis agreed, after a great deal of back-and-forth, that Chen would pay me in cash prior to the work. I no longer trusted the characters involved for obvious reasons and didn't want to spend my time and efforts for naught.

During this time I was working at my stand at the Rockefeller Center Greenmarket, located between 30 Rockefeller Plaza and the skating rink. It was blistering hot, and there was hardly a breeze. The collection of international flags that surround three sides of the rink were limp in the stagnant afternoon heat. The lunchtime crowd was thinner than usual, thanks to the oppressive heat, but I was still fielding the usual array of questions when Yuliana approached the counter for the first time. She was fair-skinned with straight blond hair, and she wore a brown skirt and a dark blue top. She reminded me of Gwyneth Paltrow, and beauty that she was (and is), I wanted to talk to her. I began by offering her a sample of the honey. I distribute a couple of thousand taster spoons' worth of honey samples during each twelve-hour market, so that part was reflexive. She didn't respond to my offer or even acknowledge that she'd heard me. I proffered the spoon again, my arm straight at the elbow, near her eye level. "Would you like to try the whipped honey?" I asked a bit louder. Still she did not glance my way. She was engrossed in the goat milk soaps on my counter. And ignoring me.

"What is it, you think you're too good for this honey?" I taunted. This caught her attention.

"I wondered who this person was who was so rude to me," she said later. "So I glanced up and intended to give a look to suggest *how dare you even speak to me?* But when I did, my glance was met by a pair of beautiful eyes. I remember that they were green. They were devilish and seductive, and accompanied by a smile that instantly melted my heart. I have never wanted to be without him since."

I have heard this story over many a dinner in response to the question "How did you two meet?" First off, I do not have green eyes. But, honey bees see colors differently from humans. For instance, honey bees cannot see red, but they can see ultraviolet and can detect polarized light. They have five eyes, two of which are compound with thousands of lenses, though the three simple eyes cannot help in detecting intruders. With her comparatively inadequate two eyes, Yuliana seems to see colors differently from other *Homo sapiens.* I can't explain why this is, but I have many examples. She argues vehemently about whether something is brown or green, and seems wholly satisfied, apparently, to live in a world with brown grass and green dirt so long as in the end I break down and agree that she is right.

In her native language there is no one word for blue—but one for light blue and one for dark blue. In Japanese, the word for blue and green is the same, at least when discussing traffic lights or seaweed. So perception is skewed no matter what.

Also, nothing about our exchange indicated that she experienced the lovestruck rapture she now claims to have been swept away by that first fateful afternoon. At that time she was quite reserved and cautious, and I had no idea that she was "contemplating the best strategy" to move things along while not demonstrating her eagerness to get to know me better. She probably overplayed that card, as she seemed positively disinterested, even bored.

Judging by her soft accent, she had clearly not been born in the United States. I asked her where she grew up. "You haven't heard of it," she said, sounding like even uttering these few words were well beneath her. In fact, she addressed her remarks more to the bars of soap than to me.

Unfazed, I retorted, "I've probably been there."

She laughed for the first time. It was genuine but brief. "I doubt it." This former granddaughter of Lenin would have made her people proud with her now fully resumed Soviet demeanor. Despite the heat she made me wish for a *ushanka*.*

"Let's clear this up now. Where is it?"

Silence.

"I can probably guess it."

Now she looked at me directly. I studied her blue-gray eyes—which actually *are* blue-gray—and saw them narrow.

---

* Usually made from sheepskin, or rabbit or muskrat fur, but not limited to those animals, a *ushanka* is a hat with ear flaps that can either be tied atop the hat or flipped down and tied under the chin. It protects much of the head, face, and jaw from the cold. The word *ushanka* derives from *ushi* which means "ears" in Russian. And before anyone gets cross with me, yes, I realize there is a world of difference between Ukraine and Russia. But this book isn't about the varieties of headgear found in the former Soviet empire, so please, allow a little room for poetry here.

Something akin to a smile started to form on her face. Then her expression reassembled itself into that blank look that I have since come to know and fear, so well and so much.

"There is no way you can know it," she insisted, now appearing to be mildly interested in what I had to say.

"If I guess it correctly, you have to give me your telephone number. If I am wrong, I will let you take me out for a drink," I bargained.

She looked at me confused. "But . . ." And then she laughed and started to walk away. And stopped again. She appeared to take a deep interest in the bottles of pollen on the counter, picking them up, studying them, and putting them down again. Since it was lunchtime, the busiest hour of the day at the market, there were at least a dozen other people lingering around the stand, but I hardly saw or heard them anymore.

"Excuse me," one called to me. I held up a finger but kept looking at the intriguing woman in front of me. "Okay. If I guess correctly then you give me your telephone number. If I am wrong I will stop asking."

"Go ahead," she said, in what was more like an exhalation.

I stared at her. I thought about her accent. It sounded to me like she'd grown up in the Soviet Union but not Russia. I looked at her features. They seemed to have something Polish in them. I thought about who I usually ran into in Manhattan, where there is no shortage of not only Russian people but immigrants from former Soviet Republics—Belarus, Ukraine, Azerbaijan, and so on—who might fit that mold. Then I took a guess.

"L'viv," I said, careful to use the Ukrainian pronunciation and not the Russian "L'vov."

Later she told me that when I correctly guessed her home city without any hints at all beyond what I could observe, she thought to herself, "This man is not just a simple farmer." And to make a long story short, we now live in a cramped one-bedroom apartment in Manhattan raising two boys. In the apartment above us live her identical twin sister and *her* husband, who is also named Andrew, and their two kids. The sisters' doting mother is there nearly every day and wanders between apartments at will, cleaning, cooking, accessing our apartment with a key that she claims she does not have. Living under these conditions is a peculiar kind of torture.

Back in Queens, with a cash donation to the New York City Beekeepers Association in hand as payment from Chen via Garis—Garis had come in person to the farmers' market at Union Square to give me the money, and to have me sign a receipt—I took several NYCBA volunteers with me to close up and strap the hives late one Thursday evening in August. Screening in beehives is a job that has to be done at night when the bees are all, or mostly all, inside the hives. We crossed the East River, donned our veils, and descended on the Corona apiary.

When we arrived on the scene at Chen's house, there was a cluster of reporters and police swarming the house and blocking part of the street while simultaneously keeping their distance from the bees. Word had gotten out that something was happening at the home of the bee hoarder. I spoke to the police captain and let him know that we had been hired by the owner to seal the hives so that the bees could be removed the following morning. The realtor interpreted this and Chen confirmed. We were given permission, with great enthusiasm on behalf of the captain, to go in and take care of business.

Mrs. Garis, the realtor, was wearing a scarf over her head and dark sunglasses, though it was an inky, moonless evening, and seemed to take great displeasure in the attention the property was receiving. The fact that the place was teeming with nearly as many reporters as bees was comical to me. Of course, none of the reporters had bee veils, so they were self-sequestered quite safely away from our activity and the clouds of thousands of confused and increasingly angry bees. Tom Wilk, Duncan Watwood, and a couple of others and I plodded ahead, screening and stapling the hive entrances, wrapping orange crank straps around the hive bodies, and sealing whatever cracks we could find—and there were many in these neglected hives—with duct tape, screens, and staples.

We had brought one fellow with us who had a particularly valuable role to play. A California native, Adam Johnson is an athletic, handsome, blond-haired urban beekeeper, and a former Mormon missionary who is fluent in Mandarin. He was able to position himself near Chen and Mrs. Garis for the express purpose of listening to their conversation so that we might understand what was actually going on.

The bees, for their part, were not cooperating in terms of staying in their hives. They were bearding, a descriptive term for what happens when thousands of bees hang on one another and drip down from the bottom of the hive, like a beard. They do this when it's very hot inside of the hive and there's an immediate need to bring down the temperature. It would have been easier to stuff squeezed-out toothpaste back into a tube than to get all of those bees back into their hives, so these displaced bees had no place to go except to fly around, inspect the lights on the police cars, and conduct their own interviews with the local reporters.

We plodded on. My phone kept ringing and texts kept coming, but I could not readily answer or check it with my thick calfskin beekeeping gloves. When I finally got to it, I saw that my friend Tony Bees was en route to the scene, having been contacted by his superiors at NYPD. Of course they'd called him at my prodding, and my call had been made as a result of his prodding me to prod them.

Upon further inspection of the hives, we discovered some definite problems with the bees. The hives were not in the same condition they had been even a week prior. Then, they had been heavy and full of honey, each with two tiers of healthy bees. Now some of the hives, even with two deeps, weighed only forty pounds at best, when they should have weighed 120 pounds at a bare minimum. Some quick inspections revealed what we feared—Chen had, in anticipation of the hives being sold off the next morning, stripped them not only of excess honey, but of *all* honey. The bees had absolutely nothing to eat, no stores whatsoever, and since it was the dearth period they had no real hope of gathering enough food to carry them until October, let alone until March. Chen had proven himself to be as underhanded about selling the hives as he had been about unloading his diluted honey.

Tony arrived and pleasantries were exchanged in the unpleasant circumstances. It was late and the air was thick with displaced bees, and in spite of the disagreeableness of the whole situation, there was a bit of a carnival atmosphere, what with the huge crowd of neighbors, police officers, and reporters. Tony dug into a few of the hives; he, too, found them to not only be void of stores, but to be diseased. He spoke with the captain of the local precinct on the scene. They agreed that the hives should be removed immediately. This was something I had not anticipated and I was not prepared for it.

Tony would not budge on this issue. He declared the hives confiscated, called in a truck, and at about three A.M., we started loading the fifty or so beehives onto the back of an NYPD vehicle. At this point, I was informed that the buyer, having been apprised of the current state of the colonies, had backed out, so I didn't feel awful about the confiscation, not that it was my call to make. Chen and his daughters, along with Mrs. Garis, openly expressed their frustration and anger. Adam, silently eavesdropping, was able to fill us in on their machinations and outright lies, so, knowing what I did, I had no sympathy for them.

The crowd soon cleared out, and only Tony and I were left to load the hives. Duncan, Tom, and Adam had, quite sensibly, departed into the night, deservedly so after many, many hours of work screening in the fifty or so beehives. Most of the police had left, too, save for four, sitting in their patrol cars with the windows rolled up and air conditioners blasting. Only a handful of reporters remained. They wanted to see where the bees were going. Tony and I discussed it. Where in New York City could millions of bees be relocated, with no notice, in the middle of a muggy August night? We tossed around the idea of Floyd Bennett Field, an old church parking lot in the Bronx, or even a discreet area in Central Park. The problem was that no one could be reached at that hour to give the okay to any of these locales.

"What about your place in Connecticut?" Tony finally asked as dawn approached.

"What about *your* place in Rego Park?" I countered.

First of all, my farm is the smallest farm in the Nutmeg State, measuring less than a quarter of an acre. It also has an 1868 farmhouse with no insulation and faulty heating, and a two-car garage so stuffed with beekeeping equipment there is no room for my trucks, van, or forklift. There is even an old yellow school bus that is, hopefully not only in my dreams, one day going to be fully converted into a recreational vehicle. With all of that, not to mention my own considerable number of beehives, I could not imagine where the bees would fit. "What about the driveway? Just for a day or two?" Tony asked. In my state of severe sleep deprivation it made sense to me, so off we went.

Accompanied by a police escort with lights flashing, Tony and I and millions of bees proceeded from Corona, Queens, up Interstate 95 to Norwalk, Connecticut. When we arrived an hour later, we offloaded the bees onto the driveway, and I immediately headed back to New York City in order to reach Rockefeller Center by six A.M. to set up for the market. I watched the sunrise through heavy-lidded eyes and tried desperately to stay awake, nodding off here and there and swerving dangerously. I was so tired that I did not immediately notice that I was headed to the wrong side of the George Washington Bridge and into New Jersey, costing me a half an hour delay as I went back and forth across the Hudson River.

Staying awake was made a little easier by the fact that for almost the entire drive I was listening to the story of the bee confiscation unfold on 1010 WINS radio, one of New York City's popular all-news stations. In my drained state, it was amusing to hear quotes from Tom, Adam, myself, and others just a few hours after the incident, which was described on air as a "bee bust," "a sting operation," and a real "buzzkill," among other clever turns of phrase.

It was, in a strange way, a fun and satisfying night. All in all, we accomplished a lot. Namely, some NYCBA members gained a ton of experience; for a few of them, it was their first ever beehive-moving endeavor. Next, we were able to provide a public service to the people of Corona, Queens, working in conjunction with New York's finest to rid a neighborhood of a potentially dangerous situation—unsafe and unfair to neighbors, and hazardous to the health of legal urban beekeeping.

It did not have to play out the way it did, though. Weeks had elapsed between the time we first met Chen and the night we took the hives away. During our initial meeting, we had realized that his bees were not registered. We'd explained to him how to register them, and even supplied him with the paperwork. He told us that he'd started with one beehive, but because of the good weather, they had multiplied to about fifty hives. This was an obvious lie. If true, it would make him the greatest beekeeper in the history of the world, who'd single-handedly solved the problem of colony collapse disorder.

Chen claimed to have been a professional beekeeper back in China. Doubtful but possible. Regardless, he essentially wanted to unload all of his tainted honey; get rid of the bees at a profit after having stripped them of their ability to survive the winter; and sell his house. It later turned out that he harbored six more beehives in the back of his restaurant in Astoria. Well, he initially said it was his restaurant, but then he said he was just a cook there. He changed his story every time he was asked.

According to Adam, our undercover interpreter, it had occurred to Chen that I was responsible for the buyer backing out and his beehives being confiscated. This was partially true. I did inform the buyer of the depleted state of the hives, and he changed his mind based on that. But I certainly did not control the strategy of the New York Police Department, nor did I dictate marching orders to the overwhelming media presence there that evening. In the days that followed I half expected some shady triad characters to make an appearance on my doorstep and demand retribution or extort payment. The evening of the confiscation I told Chen, through an interpreter, that none of what transpired would have happened if he'd dealt fairly with the honey and the hives, which was true. Had he dealt honestly, we could have accommodated his needs. Instead, karma stepped in and kicked him in the balls.

This story had legs, as they say. The next day, which was really the same day since we'd worked all night, I stood at my market booth at Rockefeller Center, so totally exhausted that I was fit to pass out. In fact, I took a catnap on the slate ground under my table for two hours. Then I was interviewed by at least eight different television networks, in three different languages, about the "un-bee-lievable" event. NBC.com's "Three Million Bees Seized from Queens Man's Home" was one of nearly two hundred media accounts that ran in the twenty-four hours following the incident. Several follow-up stories appeared over the next week, with increasingly discordant protests from Chen and the Garis pair. Never once did those parties seem to inquire as to the health or location of the honey bees, though.

And what of the bees? I held on to them for two days while I tried to get the NYPD, the Department of Health, and every other authority that I could think of to collect them. No one responded. Tony's answer was "I don't know, bro!" Since I could not personally care for them, and frankly did not want them, I gave all of the beehives away to foster apiaries. I placed an ad on the NYCBA Facebook page. I contacted beekeepers in Connecticut. Strangers read about the bees and reached out from as far away as California and Kentucky. Every last hive was taken by an eager beekeeper who promised to try to nourish the bees back to health.

Within another seventy-two hours, all of the bees were off my property. I kept a precise list of who took them and where they went, just in case they needed to be returned or retrieved for any reason. But there was never any follow-up. It seemed that the NYPD, the DOHMH, and Chen himself just moved on and forgot about the bees. I did not forget, though. I kept track, and I discovered that within a week about half of the colonies were dead, and none of the remaining ones appeared in good health. The fact was that the bees had been so badly treated by Chen and had such poor prospects after he removed all of their food that it was unlikely that any would survive the winter. It was a brutal, needless shame.

Still, August was not a total disaster. I had met Yuliana at the market in Rockefeller Center, midway through the month, and by the end of it, she was still tolerating me and my unconventional work hours and lifestyle. That seemed auspicious. We met frequently for drinks and dinner. She told me that she knew I was a hard worker because I eat quickly—that in Ukraine she was taught that lazy men linger over their food. I cautioned her to beware of becoming entangled with a beekeeper, as Tolstoy enjoyed beekeeping to such an extent that his wife, Sophia, questioned his lucidity. My warnings unheeded, in time, I got to know her son, now my stepson, Max, and lured him to accompany me on apiary inspection visits and to sell honey at the farmers' markets. He is a smart, handsome kid, so he was a natural there. Now he's approaching his teenage years, so he has better things to do with his time than hang around with a guy who plays with bees, alas. But who knows what the future holds?

So in spite of a month of frustration, wasted effort, and deceit, the summer was coming to a satisfying conclusion. There was a definite sweetness in the air and the promise of more good things to bloom.

SEPTEMBER

My son, eat honey, for it is good,
Yes, the honey from the comb, is sweet to your taste;
Know that wisdom is thus for your soul.

—Proverbs 24:13–14

September can be a difficult time for bees and beekeepers alike. The flowers are largely dying and the nectar is drying up. The colony as a unit is expiring, and the bees know it. There are some fall flowers, like goldenrod, bee balm, aster, and black-eyed Susan, that feed the bees, but it's nothing like the spring and early summer buffet. The bees start to feel hunger pains. This leads to contentious behavior toward all, two- and six-legged. Under stress, bees often become extremely aggressive, making it an onerous period for them. It is no picnic for the beekeepers, either.

My household is its own melting pot. We blend Judaism

and Russian Orthodoxy and season it with a soupçon of Catholicism leftovers. My several years of living in Japan also cultivated in me an appreciation for and an attachment to Buddhism. So we truly have a hodgepodge of religious influences under our roof. We celebrate a variety of holidays, which means we are celebrating quite often.

September is chock-full of Jewish holidays, including Rosh Hashanah, the Jewish New Year. With well over a million Jews, New York City has more than three times the number of Jews than there are in Jerusalem. Since Rosh Hashanah is celebrated by dipping slices of apple into honey, the week preceding the holiday is a killer time to have a honey stand in New York, and it's all hands on deck as we empty the truck of all the bottles of sweetness we can haul to the Union Square Greenmarket. It doesn't matter if one is not an especially religious Jew; apples and honey are eaten to give a symbolically sweet start to the New Year, and for Jews, the combination is as common as chocolate rabbits and colored eggs at Easter for Christians. Observing these traditions is part of the fabric of the culture—part of the group identity, if you will—even for those who feel themselves secular. For a honey hawker at Union Square, it is a sweet New Year indeed when Rosh Hashanah comes around.

A few weeks after Rosh Hashanah comes Sukkoth, which was originally an agricultural festival, like a Thanksgiving for the fall harvest. Part of the observance is to build a sukkah, a temporary hutlike structure similar to what the Jews lived in during their forty years of wandering in the wilderness after fleeing from Egypt, or what farmers would often fabricate to rest and take their meals in during harvest time. The structure is meant to be interim, frail—much like our own lives and that of the honey bee. The sukkah, unfortunately, is the source of yearly misunderstandings between humans and honey bees. This is because the upper corners of the sukkah are adorned with fresh fruits and gourds, attracting wasps. These wasps tend to infiltrate the areas where people are celebrating Sukkoth, and as often happens, bees are blamed for the misdeeds of wasps. (Though I cannot deny that some honey bees might be attracted as well. It's just that wasps are *mostly* to blame.)

Since Sukkoth is always in September or October, coinciding with autumn, the bees are hungry and desperate, as the attentive reader will have learned. September is a tough time to be a bee in general, but it is particularly tough on the drone. The drones are no longer needed inside the hive. Often when they'll return from an afternoon of exercise and socializing among the clouds with their equally idle brethren, the guard bees will refuse them entry. With nowhere to go and no food to eat, these virgins will remain outside of the colony and die from exposure or starvation. For their part, the workers are defensive and aggressive. Varroa and tracheal mites take stronger footholds within the colony. The beekeeper must be fully suited and gloved, use smoke liberally, and expect to face hostile fliers when approaching a hive.

Honey bees themselves will attack fellow colonies. Stronger hives prey upon weaker ones, overpowering the guards, raiding their food warehouses, and leaving them decimated and virtually empty. This is called robbing. The strong colonies attack weaker domiciles en masse in order to relieve them of their winter provisions. Fortunately, there are ways to prevent this from happening, or to slow it down. Small screens may be placed on the fronts of the hives, reducing the size of the entrance to make defending it easier. If the beekeeper is able to see the behavior commencing, he or she may throw a wet sheet over the weaker hive, which will deter a strong colony from robbing it, yet allow the workers from the weak hive to come and go to at least try to bring home the bacon. Technically, the members of the robbing hive could find their way in and out of the wet sheet, but they won't bother, and will move on to softer targets. But all in all it is a tough time for the honey bees, and they bring that frustration and desperation to bear in how they behave toward one another and their keepers.

At the farmers' market, we have an issue during these months with honey bees, wasps, and other insects paying too much attention to our products. Desperate for food and with a serious lack of nectar, the bees are drawn to vendors who sell any sweet items such as grapes, jams, maple syrup—and, of course, honey. On late September days that are warm, we often have massive clouds of honey bees, sometimes numbering in the thousands, clamoring to extract the nectar from within the bottles. I oftentimes feel a little sorry for these flying creatures with brains the size of sesame seeds. This is a proposition that's never going to work—as bright as they are, the bees have not yet learned how to topple over and break a bottle or even unscrew a lid. But if the bottles are even just a tad sticky with honey residue, the bees will be all the more aggressive and come in even greater numbers. If, G-d forbid, a jar of honey drops and breaks open, the honey bees will convey that message to their sisters with the waggle dance, and in short order throngs of *Apis mellifera* will be gratefully sucking that already-digested nectar into their bodies for return to their colony. Though it creates a spectacle at the market that is appreciated by some and entertaining to many, the thick cloud of dancing bees does not enhance sales.

This leads us back to the sukkah. In New York, with the population such as it is, these tipsy, temporary structures can be found all over the city for the seven to eight days of the Sukkoth celebration. Sukkahs can be made of organic materials topped with branches, palm leaves, or bamboo, and are frequently decorated with shiny ornaments, pictures, wind chimes, and, in the upper corners, dried, plastic, or even, as previously mentioned, fresh fruits and gourds. They are erected outside of synagogues, in front of kosher restaurants, and even on the backs of trucks, which are then driven from place to place to make them accessible to all the faithful.

So the fresh foods, like small pumpkins and squash, that are hung from the corners of the sukkahs attract the wasps and yes, okay, the hungry bees. And the riper the foods get, the more the insects are drawn to them. And once the ravenous insects discover that there are sweet secretions to be lapped up, they return to the hive to tell their sisters, who return in force, just as with a honey spill at the market. So I get loads of distressed calls and emails from concerned religious types when Sukkoth rolls around.

Aside from inciting fear among some of the Semitic faithful, September brings beekeepers and honey lovers out to

Rockaway Beach in Queens to celebrate the New York City Honey Fest. Annually since 2011 the New York Honey Fest has taken place along the boardwalk at Rockaway Beach. It is a great event and gets bigger every year, and is now run by my former beekeeping student and apprentice Tom Wilk.

The New York City Honey Fest is a lively celebration of the area's beekeepers, their industrious charges, and the mouthwatering honey that they produce together. A score of beekeepers set up shop on the boardwalk, and most of us sell out of everything we've brought while drinking craft beers and eating tacos and arepas from the local vendors. I usually also take a quick dip in the Atlantic Ocean just behind us.

Many New York City–based beekeepers sell their honey at the event, including Tom Wilk, of course; Ruth Harrington, mother of four and keeper of both chickens and bees; Ralph Gaeta, Astoria resident and father of two, who keeps his beehives on his roof with views of Hell Gate Bridge. All of them began their beekeeping careers by taking the course my father and I teach annually for the NYCBA, and it gives me great pleasure and a small measure of pride to see them with their own bottles of honey and their distinct labels, and mentoring up-and-coming beekeepers themselves. Aside from all of these homegrown urban beekeepers, others from New York City and the surrounding areas attend as well. Representatives from national beekeeping supply companies, makers of beeswax soaps, and purveyors of woodenware for beekeeping saturate the boardwalk, and several thousand people come out to support the industry and get sweet deals on local honey.

One year when I was there with Yuliana and a few other friends, the sun was stronger than we had expected, so we fashioned a wide-brimmed bee veil into a sun hat for Yuliana, who had braved the eighty-minute subway ride to reach me out in Far Rockaway. Our friend Mia Woolrich, a native of Australia who was living in Rockaway Beach, was helping us sell honey that day. I had known Mia for many years, since she first arrived in New York at seventeen to commence her modeling career in America. She has gone to great heights in the industry but could not be more down-to-earth. More impressive to me than her modeling success is her beekeeping skill set. After being part of the inaugural group of NYCBA apprentices, she set up her own hives out in the Rockaways and has never looked back. She once helped me tear up a brick wall in the alley behind the old CBGB club in Greenwich Village when an infestation of honey bees made a hole behind the bricks their home, terrorizing a very sensitive and special man who owned the apartment above it. "Get them out! Get them ouuuuut!" was about all he could shout between his tears. So Mia is not just another pretty face obscured by a bee veil.

But back to the Honey Fest, or, more particularly, its location. I have beehives in each of the five boroughs, but there is no honey from any neighborhood that tastes as good to me as the honey from the Rockaways. I don't know exactly why this is. The bees seem to be foraging on something unique to the area. Maybe it's the thousands of acres of untouched growth around the coast of the salty Jamaica Bay, or something in the marshes and weeds around John F. Kennedy International Airport. But the Rockaway Beach honey to me is the most divine. This divinity cannot be hurt and is possibly enhanced by the fact that some of my beehives are housed at a modest monastery called Buddhist Insights. Thanks to the benevolence of the orange-robe-wearing monks there, I have a place of refuge for my girls, and the girls assist with pollinating the monastery's flower and vegetable garden.

The Honey Fest is a lovely event to round off the season, but late summer/early fall is not all sweetness. In 2011, Hurricane Irene swept through the Eastern Seaboard, including New York City. Aside from the much more serious and sad business of killing about fifty people and causing billions of dollars' worth of damage, the storm battered the city's honey bees with its winds and rain. The storm's aftermath sparked a fierce debate between rival beekeeping groups as to who could lay claim to some displaced bees, and prompted an important epiphany for me. Namely, if I would leave the safety and security of academia and focus on bees as my only job (spoiler alert: I did leave academia).

Just prior to Irene, I had spent a few days visiting every one of my beehives, particularly concerned about those that were anywhere near the edge of a building, to make absolutely sure that they were secure. Of course, hurricanes can uproot trees and do all sorts of damage, so there was only so much I could do to guarantee a beehive would not become a 250-pound meteor during the tempest. But I did my best.

Irene hit New York City as a tropical storm with substantial winds and rain, starting in Coney Island, and creating havoc and flooding in several parts of the five boroughs. Very soon after the storm passed, I was behind the wheel of my truck, charging south down the deserted FDR Drive along the East River. It was a humid and windy morning of syrupy, damp air, and yet the sky was still pale and overcast, almost smoky, obscuring the promise of sunshine. I was on my way to Brooklyn, having been told that the storm had torn a hollowed-out branch from a tree in Fort Greene, exposing a hive of feral bees that lived inside it. Bees often house in the dead limbs of decaying trees. These colonies are the first to be exposed when a storm tears weakened limbs apart, and so this was one of many such calls. It came from Liz Dory, a beekeeping Brooklynite who had seen the broken limb herself and worried for the plight of the colony.

The Fort Greene feral bees were up for grabs to whoever could capture them. Word spread quickly in urban beekeeping circles on Facebook and Twitter, and hooded hopefuls made beelines from all over the city to claim the hive, or at least to watch the entertainment unfold. The stakes were decidedly high for some: A new hive meant new honey, and for many in the current world of urban beekeeping, in which apiarists peddle their products at weekend farmers' markets where activity resembles the chaos of the New York Stock Exchange, the general attitude is that the more bees one has at one's beck and call, the better. Many people believe that a beehive is a gold mine. Not so. A *Wall Street Journal* headline from April 2019 pretty much sums it up: NYC BEEKEEPERS ARE MAKING HONEY BUT LITTLE MONEY.

There's a well-known beekeepers' proverb that dates back to at least the 1600s: "A swarm in May is worth a load of hay; a swarm in June is worth a silver spoon; but a swarm in July is not worth a fly." My race down the FDR was on one of the last days of August, so a swarm, or in this case, a rescued feral hive, wasn't worth much. Maybe that's why in the past four hundred years no one has added a line about August or September to that rhyme. In any case, becoming the caretaker of bees this late in the year meant inheriting a problem. For no good reason, I was trying hard to be bequeathed this particular problem. But I love bees, and I love problem-solving wherever bees are involved, so while I was certainly in it for the welfare of the bees, I was also in it for the fun.

I drove past Spike Lee's studio and found a parking spot, climbed out of my Toyota Tundra, and clambered up a small wall to enter the park in Fort Greene, avoiding walking the extra fifty yards to the entrance so I could get to the hubbub as soon as possible. I came upon a group of people crowding the area beneath the tree with the broken limb, using their phones to snap photos of two masses of agitated honey bees. One mass, about half of the colony and probably twenty thousand or so, flew in a frenzied whirlwind in what looked like a dark murmuring cloud in the branches. The other mass was buzzing around the half of the limb that had snapped off and now lay on the ground, its stores of yellowish brown honeycomb clearly visible inside. Poor, confused bees. Their house had ripped in two, and they didn't know what to make of it.

Within half an hour, representatives from different beekeeping groups were there, along with unaffiliated beeks and non-beeks alike. I stood uncharacteristically silent, listening to a small group of people bickering over something they had no means to obtain. Some had no veils, no equipment, no experience. Most had no clue. And none of them could reach the bees; this was an academic exercise for them. Yet many of them seemed to feel completely entitled to the remains of the hive that clustered thirty feet overhead. Not to mention that at that point, no one had a clue where the queen was, or if indeed she had survived at all.

While I was talking with the people on the scene, a bee-keeper I didn't know approached me. Her name was Margot Dorn, and she said that she had seen the colony first and claimed it as hers. She then explained that she'd called another beekeeping group to help rescue it and emphatically, even desperately, repeated that since she had been the first to spot it, the colony was hers.

More than one beekeeper made his or her case as to how the colony should be rescued. Some ideas weren't bad, but all lacked the required equipment. Unless some of these bee-keepers sprouted a double set of wings quickly, those bees were not going anywhere. I continued to stand back and watch. After some time, realizing that she could not obtain it on her own, Dorn admitted defeat and relinquished the rights to the colony to me. Without protest from anyone or fear of being accused of stealing, I set to work.

Once Dorn abnegated her claim, I had called Tim O'Neal, who lives in Brooklyn and had assisted me from time to time over the years, and Tony Bees, partly because this was a big enough job that I would need help with equipment and support, and also because this was city property, a public park, and I naturally lacked authority to do anything as extreme as cutting tree limbs. A short spell after the call to Tony, however, Dorn returned and announced a change of heart. She was reclaiming her bees, she asserted, puffing herself up with authority.

"I wasn't going to touch the hive before you spoke with me, but at this point, I've already made some calls. So we'll see. If you get to it first, it's all yours. I won't try to stop you," I told her. Her face screwed up like she had eaten something sour, and she turned on her heel and marched away.

It was true. I wasn't going to try to stop her. If someone else could reach the bees, fine. At this point I was more interested in watching the beekeepers than in the bees themselves. I settled on a cinder block and observed as more and more beeks piled into the park and stared up, bemused and helpless, at the colony. That colony which, as I've said, had very little value. I was reminded about the story of the man who saw an apple atop a tree, claimed it for his own, and would not allow a man with a ladder to take it. The apple rotted and no one got it. Like the apple, the bees would perish now that the nights were getting cold and their home had been split in two.

"If we had . . ." "What if we . . ." "Maybe we could . . ." Plenty of harebrained hypothetical scenarios were put forth, useful only in terms of their entertainment value. I waited for at least two hours, chatting with Tim and watching the circus before me. Such things I had time for and interest in before having children stole all of my formerly free time.

A beekeeper whom I will generously call Poindexter Pants-tootight had been notified by Dorn, and he arrived at some point. One of his cronies came up to let me know that Poindexter was working on getting sections of plastic PVC pipe to make a long tube attached to a vacuum to suck out the remaining bees from above. I smiled and encouraged him to do whatever he thought was best.

"Do you think it will work?" he asked.

"Does it matter what I think?"

"So you don't think so?"

"I know it won't."

"*Yes. It. Will!*" he proclaimed, widening his eyes, willing the idea to succeed.

It never got past the talking stage.

Then Tony showed up, and together we worked out a plan to get a van with a lift bucket to take us up to the bees. Some people on the scene had been talking about scaling the tree and climbing out onto the branch. This was a reckless idea for many reasons. Everyone knew that the branch was weak and hollowed out. Even if it didn't break under the weight of a person, a fall from that height could mean serious injury or death. No hive is worth a broken spine or fractured skull or the loss of life. Of course, I've mentioned that some people, myself included, get the fever when capturing bees and do not always think straight. I will admit to having considered scaling the tree as well, until I remembered gravity.

The bucket truck idea was better. We pulled out tools from Tony's pickup and mine—a chainsaw, handsaws, ropes, our veils and gloves, and a hunting sock, which is a sort of big expandable net meant to secure the carcass of an animal like a deer. We established a perimeter with yellow hazard tape. When it was time to commence the real work, Poindexter walked off, not to be seen again that day.

I got into the bucket and ascended toward the hive, operating the controls for the lift in the bucket. I had never done it before, but it was pretty straightforward, and since I didn't tip the van over, I suppose I did it right. After a close-up inspection of the branch and colony, and a few photos of it and the scene below, I returned to earth, and Tony and I made a plan. We would cut away the excess branch, cover the colony within the branch with a hunting sock to secure the bees inside, tape that tightly, tie that section of the branch to the lift, then cut the section from the main branch. The log and the bees would then be slowly lowered to the ground, safely removed from the site, and placed into an appropriate home elsewhere to live out their short lives.

Many of the beekeepers standing by wore their beekeeping veils and suits in a way that reminded me of little kids wearing their Halloween costumes before or after Halloween; they do it just because it's fun. But at that distance, there was no need for veils. Meanwhile, the hundred-plus civilians—including old ladies, the wheelchair bound, and parents with children in strollers—were just as close to the action and without veils. That was fine, though, because at that point most of the bees were thirty feet above their heads, and the branch (and bees) at ground level had been blocked off with the yellow tape; no one could go near them and the bees were clustered on the broken limb. Still, we did not want a panic with tens of thousands of bees flying into the crowd if something were to go wrong. Like the branch falling, landing on the ground far below, and bursting open like a malevolent piñata.

With some simple lumberjack work, Tony and I cut off excess bits of the branch and then secured the hive in a hunting sock to stop the bees from flying hither and yon. Then Tony cut the limb free from the rest of the tree and gently lowered it to ground level. Once it was set on the ground, the beekeepers and other interested individuals clamored around the feral hive and took pictures of the huge, heavy log chock-full of bees.

Sensing an opportunity to possibly mend a fence that we did not ourselves break, and as a gesture of goodwill and community spirit, Tony and I decided to hand ownership of the bees over to Margot Dorn and her group, hoping it would serve as an olive branch—or at least a vibrating, rotted maple branch—of sorts. It was received by her with a small word of thanks.

Then, of course, things went wrong. With still a hundred or so civilians watching, Tim O'Neal and the new owners of the bees decided to set to work transferring the comb from the log into a portable hive box. In theory this was exactly the right thing to do, so that the bees would carry their home, food stores, and babies with them. In practice it was completely the wrong time and place to do it. Most unwisely the sock was opened right then and there, undoing our careful packaging and sending furious bees shooting out in all directions like stinging shrapnel. Tim was safely ensconced in the cozy comfort of his veil, his pale fingers nestled safely in thick beekeeping gloves pulled up to the elbows. The many dozens of bystanders were defenseless against the flying stinging ladies whose home had just been torn apart and then chain-sawed and hauled down thirty feet. Tony and I just about lost our cool when we realized that the bees were not sneaking out of the net through some missed egress, but had been purposely let out. We immediately closed it up again. It was unfortunate. Not a single person had been stung during the entire rescue operation; the same cannot be said for its aftermath.

All in all I was at Fort Greene for six hours, and off and on during that time I was chatting with a young woman named Emily Rueb, who, it came out later in the conversation, was a reporter for *The New York Times*. The whole silly story of the beehive rescue mission was published in the paper the next day under the headline IN RESCUE OF BEEHIVE EXPOSED BY HIGH WINDS, HONEY AND RANCOR. While it was slightly stinging, the article was fair and accurate. It showed a handful of New York City beekeepers behaving childishly, and I was there among them, perhaps no better. The story publicized a real rift that has always existed in the New York City beekeeping population, which may have made us appear to the rest of the beekeeping world as a bit, well, ridiculous. Rightly so I have to admit.

> Mr. Coté and Mr. [Pantstootight] had once attended beekeeping functions together. But Mr. Coté had a more ambitious plan for a professional beekeeping association and started his own group in 2008.

Well, attended same functions, yes; together, no. About the NYCBA it was true. Until we formed it, there was no association in existence in New York City that held monthly meetings, offered coursework on beekeeping, had a beekeeping apprenticeship program, or provided a bee package delivery service. I started the NYCBA to fill a need in the community, and to enjoy beekeeping with like-minded people. Prior to that, any interested beekeepers needed to go to New Jersey, Long Island, or Connecticut to find a reasonably sized and active beekeeping club. That seemed crazy in a city of millions. I knew we could hold our own.

> Because Mr. Coté's group regularly worked with the health department and the New York Police Department's Emergency Services Unit on such rescues, he was able to secure a police van with a crane, a chain saw, and the services of the Police Department's resident bee handler. Mr. Coté oversaw the rescue work.
>
> As throngs of beekeepers and the curious congregated within the thin piece of yellow caution tape roping off the area around the tree, tensions rose. And even as the wood chips were flying, the two beekeeping groups squabbled over how the rescue should be conducted and who the rightful owner of the bees was.

When we were in the midst of taking the limb down, upon seeing that the prize hive was indeed within grasp, the costumed beeks below began buzzing with delight and excitement—like a hive bubbling with activity as the weather warms. And why not? It was exciting. Bees, dizzying heights, equipment that made lots of noise, a muted hint of danger.

The *NYT* went on to reveal the wet-blanket, sour-grapes nature of Poindexter Pantstootight:

> [Mr. Pantstootight] said he tried to halt the operation on Sunday because the high winds trailing the storm added to an already potent combination of stinging insects, heights and chain saws. But when his words were not

heeded, he left the park. "There was a lot more testoster-one floating around than common sense," he said.

"That's strange of him to say, since he lacks both," my father said aloud upon reading the story.

Most important, once the bees were on the ground, it seemed to Tony and me that the neighborly thing to do was to share them. We both had more hives and bees than we really needed, and some of these hobbyist beekeepers would double their holdings with this one hive. And so, as *The New York Times* reported:

And in the end, who would claim the Fort Greene bees? A compromise, of sorts, was reached. . . .

On Monday, the comb was carefully excised from the branch and the bees were transferred to wooden frames in a procedure that involved a vacuum, serrated bread knives and rubber bands.

Ms. Dory will house the bees and, if they survive the winter, she will give half of them, in what is known as a "split," to Ms. Dorn. . . . She called Mr. Coté to thank him for efforts. Without his help, she said, her hive would not have survived.

But sometimes there are no real winners. Poorly managed, the bees died that winter.

OCTOBER

It is the honey which makes us cruel enough to ignore
the death of a bee.

—Munia Khan

The year after Hurricane Irene, Hurricane Sandy
hit. Sandy was the worst storm to affect New York
in more than three hundred years. Two days before
Sandy was set to strike the city, I drove over to Brooklyn to
meet Chase Emmons at Pier K of the Brooklyn Navy Yard.
They had been destined for a rooftop view, but for whatever
reason, the twenty beehives that I'd dropped off in March
were still on the dock, just a few feet from the water. With the
impending storm, I was concerned that these beehives—all
overwintered beehives and strong genetic specimens—were in
peril in their spot only two feet above sea level. Chase agreed
and asked me to come help move them.

The idea was that we would carry each of the hives to a nearby elevated spot to keep them from being swept away when the waters started to rise. We looked over the hives. They were two deeps, meaning that there were two levels of boxes per hive. I had stapled them together for transport back in March, and, five months later, they were still stapled. This guaranteed that the lower boxes of frames had never been inspected. It was disappointing, but not the most urgent issue facing us. Chase was looking at the lay of the land and shaking his head. "It seems like a lot of work to move them," he lamented.

And so, despite the purpose of my agreed-upon trip there, he determined that the hives would be fine exactly where they were. We were friendly enough, and I tried to dissuade him from leaving them so close to the water. "We can have it done in about an hour, probably less," I told him truthfully. There was a rise about fifteen feet away, and we could have just hauled them up there one at a time. Drone that he was, he would not budge for the bees, and so I left him and the twenty beehives on the dock and continued to secure as many of my own hives as I could before the hurricane hit. One of the bee-keeping apprentices also in the truck told me, "No way those hives won't end up floating down the East River."

Yuliana and I spent most of the evening of the storm enjoying the safety and warmth of my tenement studio apartment on Broome Street on the Lower East Side, and the old building showed every year of its century-plus age in the howling wind that night. I had loaded up two five-gallon pails with water, had a chest full of ice, and even a generator chained in the back of my nearby truck, just in case. When the electricity cut out, we enjoyed the dinner I had prepared earlier on my gas stove, drank wine by beeswax candlelight, and watched a movie on my laptop. When the battery on that died, we were content to listen to my old transistor AM radio. We even fool-hardily went for a walk when the wind was violent enough to make stoplights swing dangerously and traffic signs bend and sway; some signs were trying desperately to dislodge them-selves from the pavement in the fierce sustained gusts. It was hard to imagine my urban hives or anyone else's surviving all this, but I had secured my own as best I could, strapping some together and moving them as far as possible from the edges of roofs. There wasn't much I could do but wait it out.

At daybreak I peered out my oversized windows and looked toward the Williamsburg Bridge. In the foreground I saw the ruinous aftermath of Sandy: pockets of the city lit-tered with fallen power lines and broken tree branches, and soggy debris floating in streets flooded with murky rainwater. I drove Yuliana back uptown to her apartment across from Lincoln Center and promptly turned my truck back down-town to survey the storm damage and check out some of my apiaries in Manhattan, Brooklyn, and Queens.

I found that my beehives were all fine, thanks to a combi-nation of preparation and, more important but harder to guarantee, good luck. The Bridge Cafe, on the corner of Water and Dover streets in the Financial District, was now flooded with deep water. The beehives were on the rooftop—the site of the memorable *Cake Boss* filming—where I had left them. Sadly, the business would never recover. Restaurant owner Adam Weprin eventually sold the historic building.

Instead of moving them to nearby safety, Chase left the

Grange beehives on the docks and placed cinder blocks and sandbags on top of them—which did nothing but help drown the bees faster. I was saddened but not surprised.

I received some notifications about bee boxes found in lower Manhattan and in Queens that were identifiable as those swept-away Grange boxes; the hives had broken apart and been carried off into the East River and spread over three boroughs of New York. But Chase didn't let the incident get him down. As big a tragedy as Sandy was to those who lost everything—their livelihoods, their homes, or even their lives—Chase started an online campaign blaming the terrible storm for killing his bees. He quickly raised over $22,000 to replace what had been about $6,000 worth of bees and equipment that he had purchased from me that spring. He even issued a press release that hit all the right notes to open purse strings, resulting in a story that appeared in several publications.

Edible Manhattan reported, "This is a tale of perseverance. One colony managed to escape their hive, and someone at Brooklyn Grange quickly built a new hive to protect the survivors. 'We are not sure yet if there is a queen with them, but if there is, we're going to do everything we can to ensure they survive the winter. Hurricane Hive, we got your back,' wrote Chase Emmons, chief beekeeper." I was impressed, in a stomach-churning sort of way, but mostly outraged that Chase was able to spin such willful negligence into gold with some media magic. But it made the gullible all the more willing to pony up dough, apparently. Most enraging to me, aside from the needless loss of honey bee life, was this bit in the *Brooklyn Paper:* "Emmons knew his hives were at risk before the storm struck, but relocating such a huge quantity of stinging insects is no small task. 'There was little we could do without a Herculean effort,' he said."

This opportunism aside, luckily, the beekeeping community overall did not suffer horrific losses, and of course our losses were nothing compared to the human toll from Superstorm Sandy.

Not all bee tales are so sour. There are successful rescue missions, too. Back in my Connecticut hometown, settled in 1649, the Norwalk Green Historic District is located more or less in the center of the city. There is the Norwalk Green, which includes a World War I memorial and a large white gazebo on a neatly manicured and lush lawn; the space has been used for public events for more than a century and a half. Facing the green are both St. Paul's Episcopal Church and the First Congregational Church. Nearby is the Norwalk Historical Society and Mill Hill Historic Park, with a cemetery containing graves older than Norwalk itself.

My old pal Anna Veccia worked part-time at the historical society and would sometimes take her lunch on the green, enjoying something she had brought from home or partaking from one of the few food trucks that parked alongside the area. These trucks sold inexpensive Mexican food and catered almost exclusively to the Central and South American day laborers toiling in and around the area doing landscaping work. On one particular day, sitting on a bench and slowly eating her sandwich, she happened to look up at the steeple of the First Congregational Church and notice something large affixed to a spot about one hundred feet up just under an awning.

I wish I could write that a few bees flew down to her and that one settled on her knee, another on the bench beside her, and a third hovered to get her attention, whereupon they drew her line of vision upward and she spied a forlorn colony far above in the sky. Alas, it was less dramatic. Anna was simply enjoying a light snack in the slight chill of early autumn and noted what looked like a colony of honey bees that had set up shop in the great outdoors. She roused herself off the bench and walked across the street and up the church steps to take a closer gander at the sight. Upon this further inspection, she determined it was indeed a colony of honey bees, and it turned out to be the largest feral colony I had ever seen up close and in person at this latitude. It was three feet long and two feet wide, with huge, deep combs. It was enormous and beautiful and wholly out of place out there in the open on that 1848 tall, white southern New England steeple. Anna told me about it, and I planned to check it out but dragged my feet a bit.

Until a couple of years ago I did not have a website, nor a business card, and I still do not have a business telephone number. So when people managed to get hold of me they either had tried very hard or were Norwalkers who knew someone in my family. Which means that even though I had not lived with my mother since I was a teenager, people will sometimes call her, after trying a few other of the Norwalk-listed Coté numbers, in an attempt to track me down.

I later received an email from my mother saying she'd had a call from Norwalker Lyn Detroy, and it "had something to do with the Congregational Church bees." Lyn reported that there was a concern that the bees making their home on the steeple might start dive-bombing parishioners and interfering with weddings. Realizing that this must be the colony that Anna had spotted, and always happy to do something new and exciting in the bee realm, my father and I visited some of his former colleagues at the Norwalk Fire Department. We asked to borrow a fire truck with an aerial ladder to get up to the colony. As my father had been a well-respected Norwalk fire lieutenant for decades, the loan was easily arranged.

The first day that we could get to the church just happened to be the same day that a one o'clock wedding was being held there, so we knew we had to work quickly and carefully. The morning of the rescue, I woke up in Manhattan early, drove an hour up to Norwalk, and met my father, Anna, and a bunch of family members who would act as witnesses and cheerleaders near the green just as the sun was about to rise. Our plan was to grab as many of the bees as possible before the foragers started off for their workday.

Unlike a routine inspection performed when half of the colony is out foraging, this job was a little rough in that the whole colony was home, just as we had hoped they would be. While it was good that we could theoretically get all the bees, it was difficult in that we had to deal with tens of thousands of foragers, who jealously defended their home as soon as we started dismantling it. Our approach was to cut the comb sections off the steeple, pass them into waiting containers, and repeat until there was no more hive on the church. This worked well but for the fact that the colony was much larger than anticipated. In fact, it had looked as small from the ground as our friends and family now looked from up high, so we soon ran out of room in our containers.

We didn't want to descend back to earth to get more because we didn't have room in our small working area, nor did we wish to lower the thirty-five thousand recently evicted bees and leave them down among Norwalk's bravest and our kin. The comb, which was heavily laden with honey and brood, is made of thin beeswax, so it was not strong to begin with. Even though the hexagon is the strongest shape found in nature, when the comb is turned on its side and dislodged from its original home, it becomes very weak. We tried to avoid breaking any, but it was a big job made more difficult by a myriad of stings—in spite of our full beekeeping gear and thick calfskin gloves—and the dizzying heights.

We ended up making a sticky, messy job of it, but we got it all. The only thing that remained was a dark stain on the underside of the steeple where the honeycomb was once firmly attached, and a few years later that was touched up with white paint. In their own literal way and though I did not realize it at the time, these bees brought me closer to G-d than I had been in many years, or ever expected to be. Later, after transferring it all into a traditional beehive, we relocated the comb and colony to an inner-city community garden in Bedford-Stuyvesant as part of a local Bees Without Borders project.

The wedding went off without a hitch, with no one getting a bee in their bonnet.

Afterward, my father and I cleaned up, washed off, and got on a plane for China.

I had been invited by the Global Center for Hospitality Management, which is part of the New York Institute of Technology, to give a talk on urban beekeeping in Hangzhou, China, about forty-five minutes southeast of Shanghai via the fastest possible train, as part of the First International Forum on Sustainable Development of Hospitality, Education, and Industry. It was a huge and exciting conference, and I fully expected it to live up to its very long title.

A year earlier, I had been in the office of Dr. Robert Koenig, the assistant dean of hospitality studies at NYIT, to talk about placing beehives on the roof of the school. I immediately found him affable, polite, and brimming with boundless energy and enthusiasm. His office was filled with awards, certificates, photographs, and newspaper articles from his decades in the hospitality business, and a huge wine collection.

As we talked about the possibility of placing beehives on the roof, I consulted with a friend I had brought along with me, Dr. Fumio Sakamoto (坂本 文夫). A fellow beekeeper and professor from Kyoto Gakuen University, Dr. Sakamoto was visiting from Japan for a few days and staying at the New York Hilton Midtown, where I maintained an apiary of six beehives. We had spent the day hopping from one rooftop apiary to the next as I returned the favor and courtesy he had extended to me earlier in the year, when my father and I had tagged along with him visiting rooftop apiaries in Kyoto and its outskirts. He had also taken me to one of the three shops in Kyoto that specialize in selling pure honey. Pure, unadulterated honey is a rarity in Japan, even more so than in the United States. More than 94 percent of the honey sold in that country is from foreign sources, usually China, and, as earlier mentioned, that so-called honey is often not actual honey but rather genuine honey heavily diluted with simple syrup. Dreadful stuff.

Dr. Sakamoto is a foremost authority on honey bees, and I was familiar with his work prior to being lucky enough to meet him. He escorted my father and me to his honey bee laboratory at the university, to his office (which looked like that of a mad scientist with a bee fixation, which is pretty much what it was), and not only to his own apiaries but to those of his friends. One beekeeper, a fellow about fifty years old, was using a hive tool that was different from any I had ever seen before. I openly admired it, and before I could stop him, the man insisted that I accept it as a gift. I insisted that he keep it. He was not going to back down, so in the end we compromised; we would trade hive tools, mine for his. Then he told me that his father, who has since passed on, had given him the unusually shaped tool. So we shared a reflective moment in the meadow in which we were standing, in the tall grass punctuated by yellow flowers, and spoke gently about all that our fathers had done for us in beekeeping and in our lives.

Then my father, who had been playfully hiding in the waist-high weeds and flowers, stood up to reveal his presence. He pantomimed growing up out of the earth as a beautiful flower himself, extending his arms as if he were blossoming. The man and I laughed at him. We had been speaking in Japanese so there was no way that Norm could have understood the conversation, even if he could have heard it—which he couldn't have, given that he is largely deaf from too many years of fire truck alarms and sirens. But his timing, as always, was impeccable.

Back at NYIT, Dr. Koenig explained that he wanted his

students to do something hands-on that was outside of the usual curriculum, and he and Dr. Sakamoto and I talked about urban beekeeping as a possibility. During the course of the conversation, Robert, as he insisted on being called, asked if I would be willing to talk about urban beekeeping and rooftop farming at a conference being planned for the following October in Hangzhou, China. Marco Polo had called Hangzhou the City of Heaven; of course I accepted. And, naturally, I asked if I could bring an assistant. This is how my father and I ended up flying to Hong Kong one late October morning and, after spending a day in Kowloon and on Hong Kong Island, catching a connection to Hangzhou.

In Hong Kong I had hoped to meet up again with urban beekeeper Michael Leung. We had met in New York City several years prior when the NYCBA hosted him as a speaker. Our schedules did not quite mesh, but we did find time for a phone conversation about beekeeping practices in mainland China and our mutual passion for maintaining beehives. It's nice to know that in nearly any major city in the world, I have a counterpart with whom I can compare notes about beekeeping—Marie Laure Legroux in Paris; Dale Gibson in London; Torbjørn Andersen in Reykjavik; Nils Simon in Berlin; Kaethe Burt-O'Dea in Dublin, Kazuo Takayasu in Tokyo, and Isamu Nishimura in Kyoto. Good beeks are all over.

I knew very little about China and even less about beekeeping in China. What little I knew included that in 2014, a fellow named Ruan Liangming had sixty caged queen bees strapped to his body, resulting in about 140 lbs. of honey bees landing on him, awarding him the Guinness World Record for the "heaviest mantle of bees" ever worn. And that in 2009 a couple of beekeepers, Li Wenhua and Yan Hongxia, decided to be likewise covered in honey bees while taking their marriage vows. In other words, at least I knew that beekeepers in China were likely to be as obsessed with bees and beekeeping as most any other apiarists.

More seriously, in China, with an incredibly large population to feed, cities increasing in size, and food becoming more scarce with land disappearing into urban sprawl, beekeeping is becoming more and more of a challenge. Heavy pesticide use continues to present a tremendous challenge for the little honey bee, which doesn't fare well in the hostile environment of chemicals. Colonies wither and die. With honey bees on the decline, it's imperative to find an alternative for the pollination services that the honey bee would normally provide. And the Chinese have found one: hand pollination.

In Hanyuan County, for example, the self-described "World's Pear Capital," unchecked pesticide usage has all but eliminated honey bees. Pear tree flowers, which are not especially attractive to bees in the best of times, are consequently pollinated by hand by men and women who climb ladders, scale trees, and transfer pollen, which they carry in little jars, to the stamens of the flowers via long sticks with feathers attached to the end. They dip the feathers into the jar of pollen, taken from male blossoms, and they dust pollen into the bud to stimulate the growth of the fruit in the female flowers. In the absence of insect pollination and without human-assisted pollination, the trees would not bear fruit.

The problem of *Apis* species decline is not unique to China, though in some places there, it is extreme. Bee diversity has been a fact of life for millions of years on planet Earth, but it's diminishing all over the world at an alarming rate. This, of course, includes thousands of native bees in addition to the honey bee that we know and love. The Macropis cuckoo bee, best known as the bee home invader that will take over another bee's nest and lay her eggs in it, is at the top of the endangered list in North America. The Great Plains are now nearly void of the oversized sunflower leafcutter bee; the three-lobe snouted sweet potato bee from the East of North America is also dying out. Sadly, there are thousands more examples; perhaps as many as one in four types of *Apis* species are under threat of extinction.

Since bees may be considered the canary in the coal mine of our ecosystem, this does not bode well. It is widely reported that Albert Einstein said, "If the bee disappears from the surface of the earth, man would have no more than four years to live," but he didn't really say it. It was said by Maurice Maeterlinck in 1901, and it's not true. Still, it *is* true that honey bees contribute pollinating services to food for billions of people, and without them, we would have a much blander diet. Unless, that is, we all start stapling chicken feathers onto the ends of long sticks and handpollinating our favorite fruits and vegetables. So while the honey bee itself may not be in as much danger as many people believe, it is true that in general terms, bees of all stripes are under varying levels of threat from habitat loss, pesticide mismanagement, and climate change, and the honey bee specifically is greatly harmed by these issues. The diminishing population of *Apis* species is a complex issue, but one chief takeaway is that the lack of care for the ecosystem as a whole has devastating consequences for our pollinators, and this goes beyond bees and into butterflies, birds, and other creatures. But I digress.

Robert had arranged for a car to take my father and me to the hotel when we arrived in Hangzhou. We had imagined we would be treated well, but we had no idea what a luxury event awaited us. The conference was massive, with several hundred attendees from all over the world. At least two dozen delegates came from New York, and there were others from many other locations. The hotel room my father and I shared (New England beekeepers, we are pragmatic even when someone else is footing the bill) was beautiful and afforded us views of the city that might have been even more spectacular if not for the poor air quality. Everywhere one looked there was construction. The food was spectacular, though unfamiliar. We are adventurous sorts—my father still remembers eating mopane worms, which are actually large caterpillars, in Zimbabwe. I've had my fair share of unusual food and drink in my travels around the world. But nothing could prepare us for the delicacies of Hangzhou cuisine. Duck tongue. Or, the entire duck head. Still-squirming live seafood. Spicy river snails. Pickled chicken feet. And— though I don't dine on swine—we were offered pig ears and a great deal of pork options. Each item I tried proved more delectable than the last. It was hard to settle on a favorite.

At the conference, many speakers ran over their time limit and my original half-hour presentation was reduced to a mere seven minutes. I gave a one-sentence introduction in Cantonese that garnered rousing applause—no doubt mere politeness. Then I switched to English at double the normal speed, flipping through my slides perhaps too quickly to be fully or even partially absorbed. The interpreter had no chance to keep up, I fear. Thank goodness for the photos and videos of the bees, which needed no words and carried my message of the beauty of the small creatures, even in—or especially in—an urban environment. So in seven minutes I earned our keep for the trip.

The conference was energizing on many levels, and I met several interesting and accomplished people, but for my father and me, the real fun began at its conclusion. My fervor is perpetually with the bees, and so to that end, I had my good friend Mio, who had spent a year in China, help me find beekeepers in and around Hangzhou. As fate would have it, Hannah Sng Baek, my trusty worker from the farmers' market, was studying in Korea for a year, and was able to make the trip down to meet us. So we three beekeepers met in the lobby of our hotel the day after the conference, were greeted by our guide, LingLing, and headed off to get to know some of the four-winged, six-legged flying Chinese.

In addition to LingLing, we were accompanied by a driver, who drove us for a long time, it seemed, never reaching the end of the massive city. But eventually, after a few stops that included an ancient pharmacy and some odd yet tasty snacks here and there, we arrived at our first and primary destination for the day: the clinic of medical doctor and apitherapist Dr. Zhang.

Apitherapy is the practice of using products from the beehive, specifically the venom of honey bees introduced into the body via a sting, to cure ailments. It is practiced worldwide, and those who support its use claim it helps cure or lessen the pain of multiple sclerosis and arthritis, among other things. The truth is that the results of many medical studies do not support apitherapy. Still, the practice has persisted since the time of the ancient Egyptians and Greeks, who used it in their traditional medical practices. A moderately more modern study comes from an Austrian physician who published a monograph in 1888 entitled "About a Peculiar Connection Between the Bee Stings and Rheumatism." Decades later, in the 1930s, a Hungarian doctor minted the term "bee venom therapy." (Hungary has a very strong beekeeping tradition.)

Dr. Zhang is the only medical doctor licensed to practice apitherapy in Hangzhou, a city of about twenty million souls. Not that each of them was clamoring to have their ills cured with potentially painful bee stings, but still, he will never be short of patients. There is a lot of competition for medical doctors in China, he told us, and working with honey bees in his treatment regimen helps him to stand out. He completed his medical training in 1972, and began offering apitherapy treatments in 1990, after being asked about it by many of his patients. He conducted his own research and eventually began to keep his own colonies of honey bees since there was no reliable supplier of bee venom or honey bees in the area. He then started practicing on himself in order to learn how best to handle the honey bees and to know firsthand the effects of the stings.

"The first time I tried to sting myself, it was very difficult, and I was having a hard time catching and taking the bee," he told us through an interpreter. "Then I could not get the bee to sting. Then I saw that I was actually at that point holding on to a fly and not a bee." We all laughed. Dr. Zhang was a pleasant man who seemed to genuinely want to better his patients' lives. And he loved his bees, too. A sad and unavoidable irony was that in performing apitherapy, he sacrificed the bees' lives.

Dr. Zhang practices traditional Chinese medicine, and as part of that practice, is skilled in acupuncture. He easily transitioned the principles of acupuncture into his apitherapy treatments, using the same meridians to apply stings where he once applied metal needles. Dr. Zhang believes that "being a beekeeper makes you a healthier person because you naturally get stung a lot, eat a lot of bee products, live a healthier outdoor lifestyle, and live longer than average." It may well be true that the Calvinistic approach to toil inherent to beekeeping is a secret to longevity. Waldo McBurney (1902–2009) of Kansas lived to be 106 years old and worked as a beekeeper until nearly the end of his life. He was recognized as the oldest working person in the United States at the time. In any case, Dr. Zhang is currently training his daughter, also a physician, to administer apitherapy as well, to continue his lineage.

Dr. Zhang maintains colonies of beehives atop the fifth floor of the hospital where he works; his wife helps to maintain them. He also runs a small private clinic nearby, which is where we found ourselves that morning. He spoke no English. The six weeks of private Mandarin lessons I took prior to my trip had done me no good at all. We did have an interpreter with us, but it was difficult for her to keep up with three excited beekeepers asking questions and the specialized vocabulary involved. Still, we managed to take in a lot and loved it. We could tell that our host was excited to have a small foreign delegation visiting him in his clinic, too.

We watched as Dr. Zhang treated several patients in his clinic. Arthritis was a common ailment among his elderly clientele, and most of his practice was devoted to assisting those afflicted. One woman was so crippled by arthritis her hands were like claws. She visited the clinic three times a week to receive ten to fifteen stings per hand, which she said allowed her some limited use of them. Without it, she told us, they were totally useless and much more painful.

Some of the treatments, Dr. Zhang freely admitted, were merely for the relative comfort of the patient and not expected to bring about a full recovery. Treatments of multiple sclerosis are palliative, he said, as is the treatment of herniated discs. But arthritis is among the conditions that Dr. Zhang is convinced can be cured in most cases, and he believed that this woman would regain near full use of her hands in time. Dr. Zhang is also looking into how apitherapy can help nerve issues, and possibly tumors or even frostbite. He's nothing if not ambitious.

My father suffers from chronic neck pain from an injury he received while on the job as a firefighter. We also decided we needed our own guinea pig to test out the efficacy of apitherapy, so Norm was appointed to sit in the chair. The good doctor took a small plastic bottle filled with about thirty live honey bees from the far side of his desk. It looked like an eight-ounce miniature recyclable bottle that had once held soda, and it was cut through the middle so that the entire thing could open up like Pac-Man. It closed when not handled, and it looked to me like the bees should be able to easily get out—but none seemed to. Next to it sat another plastic bottle, this one clearly a former water bottle, that held the still-living worker bees that had already sacrificed their stingers and venom sacks for the betterment of Hangzhou's apitherapy community. Though honey bees die once they sting and their abdomens tear open, their death is unfortunately not instantaneous.

Because Norm doesn't hear well, he did not fully realize that he was about to be stung when Dr. Zhang deftly opened the first bottle, extracted a single bee with a pair of tweezers, and pressed her to my father's neck in one swift motion. It wasn't quite as impressive as a martial-arts master capturing a fly from the air with a pair of chopsticks, but it was close enough. The bee stung Norm, and he almost imperceptibly jerked forward. The doctor laughed and Norm smiled, protesting that if he'd been ready, he wouldn't have moved. It is true. Neither my brother nor I react much to bee stings since the only response we ever got from my father while growing up was the sound of his laughter. Soon we learned not to give him that satisfaction. He doesn't react to a sting too often, either, so we don't get much of a chance to laugh at him for that.

Immediately post-sting, Dr. Zhang started to tap and massage the area in order to distribute the venom. He administered a few more stings, and Norm accepted them unflinchingly. Later he said his neck felt a little better. But maybe it felt better because the stinging had stopped, like how someone's head feels better when they stop pounding it against the wall.

Back in Manhattan, Yuliana was experiencing head and neck pain that she'd been unable to shake. She'd tried several different doctors and approaches, exercises, massage, and all sorts of remedies. Though Yuliana is hardly what I would describe as a strong advocate for alternative medicines, she agreed to see a friend of mine, Dr. Patrick Fratellone, a medical doctor and beekeeper who practices apitherapy right out of his office on East Fifty-seventh Street. Like Dr. Zhang in Hangzhou, Dr. Fratellone pulls bees out of his own hives and carries them to his practice in a small container. After some preliminary blood work and an hour's worth of intake questions and examination, Yuliana was ready to meet with him two weeks later to be stung.

This would be only the second time she'd ever been stung by a honey bee. The first time was atop the now-defunct Waldorf Astoria. That iconic hotel was near her office, and I would often plan my routine maintenance of that apiary around her lunch schedule. On this day I'd had an interview with a reporter from a Norwegian magazine. Yuliana waited for me a respectful distance away on the roof—respectful of the bees, so as to avoid stings. There was a sprawling raised garden there where the hotel chefs like Peter Betz cultivated small apple trees, strawberries, mint, and tomatoes, among other delectables destined for the five restaurants in the hotel.

As Yuliana wandered through the garden twenty floors above Park Avenue, she heard a screech. A bee had become entangled in the photographer's curly hair. The scared Scandinavian started shouting and flailing around wildly, almost uncontrollably, it seemed. If it weren't for the four-foot-high parapet, she surely would have become airborne and her story would have had a very different ending. I did not dare approach her myself. I was wearing a veil and jacket that had absorbed so many bee stings that day that my presence in her proximity, with bees still hovering around me, would have been pouring kerosene on a fire.

Yuliana calmly approached the reporter and dislodged the bee from her long locks, but in doing so, earned herself a sting on the finger. "No good deed goes unpunished," I told her later, which she did not find helpful or amusing, though I assured her it was both.

"I had to look like I knew what I was doing, since I was the girlfriend of a beekeeper. I had to save face, too. So I did not let on that I had been stung," Yuliana replied in her usual soft tone.

A long pause. Then she mused, "It's cheaper than Botox and the results are probably similar if the stings are in the right places."

"'Bee-tox'?" I offered helpfully, and again, unappreciated.

The photographer had refused to wear the veil I'd offered her, explaining that it would interfere with her ability to use her camera. Lots of photographers and videographers refuse the veil and end up stung. I suppose they know the risks when they take the job. Trained by my father, I have to stifle a laugh when others are stung.

The point is that Yuliana did not appear to be allergic to honey bee stings. In fact, only about 1 percent of the population is. Many people will react with localized swelling and itchiness after the initial prick and injection of venom, but a sting isn't life threatening to most people. On the other hand, saying the wrong thing to a Ukrainian woman is a surefire way to put one's life in jeopardy. The world is full of dangers.

I mentioned that Yuliana resembles Gwyneth Paltrow. In some photos, in fact, the resemblance is uncanny. With the greatest affection for both women, they both seem to have a few sandwiches missing from their respective picnic baskets. As of late, Gwyneth has been known not only as a star of stage and screen, but off-screen as an advocate for some unconventional health regimens. For instance, she has lamented the possibility that we may in fact be hurting the feelings of our glass of water by expressing negative thoughts around it, since, she says, water has feelings, too. She has also been quoted as using mugwort to steam sensitive areas of her body to rejuvenate them. Germane to our tale, along with an increasing number of Hollywood celebrities, Gwyneth purposely had honey bees sting her face for health benefits. "It's actually pretty incredible . . . but, man, is it painful," said the woman who plays Pepper Potts. But if her look-alike could handle bee stings as a cheap Botox substitute, Yuliana decided, so could she.

At the very least we knew that Yuliana was unlikely to die from anaphylactic shock from the intended apitherapy. She liked Dr. Fratellone, whom I'd first met about a decade earlier, when he gave a talk to the NYCBA on behalf of the American Apitherapy Society. I decided to accompany Yuliana to the office visit because I wanted to know exactly where and how Dr. F. would have the bees sting her, and not in the slightest bit because I enjoy seeing her in a little pain, I swear. If it proved effective, I would treat her myself as often as necessary, or maybe more for good measure, just to bring her relief from her head and neck pain and, again, not at all from any muted sense of enjoyment on my own part.

In the end, it was all up to me. Dr. Fratellone wasn't available one day when Yuliana's neck pain became too much for her to tolerate. "Just get a handful of your bees and let's try it!" she surprised me by saying one day. So I did what she said—almost. I went to a nearby apiary and managed to coax a dozen or so worker bees into a small box—it was cool outside, so this was not the easiest of tasks—which I then quickly put inside my coat for safekeeping and warmth. The box was secure but allowed airflow, and the bees could not escape. That is, so long as I closed it snugly, which of course I did not. So by the time I got back to our apartment, I had conducted accidental self-apitherapy, a new concept, on my chest and stomach. Still, I had half a dozen workers ready to make the ultimate sacrifice for Yuliana's betterment.

She lay down on the bed and flipped her long blond hair up

over her head. I felt like she was hiding her face to mask her pain, or perhaps even her tears. But I didn't waste any time by dwelling on that; one must break the shell of an egg to make an omelet. I set to work, gingerly gripping the bees one by one by gripping their wings with my fingertips, pressing them against her flesh, and thereby provoking them to sting her. Within a few minutes there were half a dozen pulsating venom sacks protruding from her neck and shoulders. There was some localized swelling and a little redness. I made sure all of the poison was in, removed the stingers, rubbed the area with a little alcohol like I had at the onset—both to disinfect the area and to distribute the venom—and she turned over and sat up.

"Are you all right?" I asked, concerned, since she had not said much to me.

"Yes," the master of understatement and subtlety responded. I thought she might sound weepy.

"It didn't hurt?"

Yuliana paused, looked at me with narrowed eyes, then chuckled. "Growing up in Soviet Union, dentist did not use anything to dull the pain before he drilled into bone. I birthed both my children with no painkiller. You think your little pets are going to . . . what . . . make me cry?" And she stood up, shook her hair back, and walked out of the room, still laughing. The former Pioneer didn't turn around when she said, "I don't know if it helped my neck pain, but was nice of you to try. Make sure your bugs are not on floor for baby to crawl on."

And that first and last treatment was the extent of her apitherapy regimen.

NOVEMBER

I don't like to hear cut and dried sermons. No—when I hear a man preach, I like to see him act as if he were fighting bees.

—Abraham Lincoln

A century ago on the family farm in Quebec, my grandmother would glide back and forth on the swings that hung from the strong boughs of the cherry trees at the rear of the property by the old stone wall. At one point or another, she and each of her thirteen siblings—the progeny of faithful Catholics, begot by one father and two consecutive mothers, and perhaps just as much the result of long winters without television—used to frolic beneath those branches near to where their father, Hector, kept about a dozen of his beehives. The children were so comfortable with the bees that they used the beehives as step stools in order to better reach the higher branches. Several of the elder siblings had their own particular perch where they would sit and feast on cherries. My grandmother's sister Jeanne Laramée told me, "The cherries were small. But we ate them. They were like the family dessert. Marie-Blanche, Marie-Ange, Jean-Maurice, Marcelle, and I, we each had our branch-seat and tasted the cherries right from the trees. When in autumn they had a little frost, they were even better. My mother made delicious jellies, *surettes*, and sweets, too, and stretched on a hunk of bread coming out of the oven, it was very good."

In the shade of those trees, the children could watch the bees fly back and forth through the air as they did much the same. The bees never bothered them at all except for two weeks in late June or early July when the fruit would ripen. Some of it was too high to reach and so it would overripen, becoming irresistible to the bees, and so for a fortnight every year the children abandoned their swings and played elsewhere. They would perhaps spend more time with the cows, sheep, or the two horses, one of which was used to till the land and one whose sole purpose, as recalled by my ninety-three-year-old Tanté Jeanne, was to take the family to church on Sunday in the buggy, after placing hot coals in iron boxes to keep the children's feet and backsides warm.

Cherries similar to the ones from my great-grandfather's trees are used in the manufacture of maraschino cherries, a business that is surprisingly large and rife with competition. The harvested cherries arrive at the factory in a dried-up state and are then reconstituted with a solution of FD&C Red 40—also known as Red Dye No. 40—sugar syrup, and a few choice ingredients to produce a plump, unnaturally red, perversely sweet cherry creation suitable for adorning ice cream sundaes.

The largest of all the maraschino cherry factories in the United States was founded in 1946 by Arthur Mondella, Sr., in Red Hook, Brooklyn, where the film *On the Waterfront* was set and the gangster Al Capone was raised. Red Hook has experienced a renaissance in the last decade. For a century it had been an underserved industrial area but is now, like much of Brooklyn, gentrifying. Through it all, Dell's Maraschino Cherries has prospered for nearly seventy years. In 2011 it was reported to be a $20 million a year business thanks to contracts with Red Lobster, TGI Fridays, and Chick-fil-A, among many other chains, some of which appropriate their sickeningly sweet cherries from Dell's not just in red but in a rainbow of colors.

Late in the season my great-grandfather's honey bees would zero in on the branches of the cherry trees unreachable to him and exploit the drippings from the splits of the overripe fleshy red fruit, or buzz curiously and helplessly around the jarred fruit in the nearby kitchen, hoping in vain to penetrate a clear glass jar, much as they do at my farmers' market stall in autumn. The honey bees of Red Hook had it much easier than their Canuck cousins. They needed only to navigate to Dikeman Street to fill their bellies with the sweet liquid that flowed down the streets and into the sewers. Very much against environmental regulations, Dell's habitually took the excess liquid used to bloat up the cherries and discharged it right into the road. It looked like blood running down the cobblestones.

In 2010, the factory was being run by Arthur Mondella, Jr., the grandson of the founder, Arthur, Sr. Arthur Mondella had a son he named Ralph, and Ralph named his son Arthur, Jr. Despite Ralph's confusion as to how that naming tradition traditionally works, the Mondellas enjoyed a prosperous business—until there was a hitch. One of Arthur, Jr.'s large clients reported to Arthur that a bee was found inside one of the jars of cherries they'd ordered. That solitary anonymous bereft-of-life bee in its urn of cherries and sweet goo is how I got to know Arthur.

But I already knew about a problem in connection with the bees and the factory. Five beekeepers in Brooklyn had independently reached out to me and reported that their honey was varying shades of red, the reddest being the closest to the factory and the hue fading as the distance between the beehive in question and the factory increased. The Department of Health had already been on the case, in its own way, by issuing nuisance violations to some of the beekeepers. One was David Selig, a beekeeping friend who had beehives in Red Hook. In the late summer of 2010 he wrote to me about the "interesting situation":

Hey Andrew! I am in an interesting situation! My Red Hook hives have drawn the nearby Maraschino cherry factory's ire! I got a warning from DOH telling me that they are issuing me a violation for the bees since they are a nuisance to my neighbors! My honey is red like cranberry juice and carries red dye #40; had a sample tested. The cherry factory uses outdoor space and has open bins of high fructose corn syrup w/dye marinating the cherries. Lots of junk must get into the juice besides bees! The factory is sloppy and pours lots of syrup into the rain drain on the street corner which also feeds my bees. There are 9 other hives (including 4 on governor's island) collecting red "honey" but are now feeding sugar water at the hives to keep bees from cherry juice. I'm sharing this in part for you to know the story but also to seek your advice.

I also heard from Tim O'Neal, who currently teaches biology at New Voices Middle School in Brooklyn, and who has been beekeeping since he was in middle school himself. Despite what you've read heretofore about some of his bee-keeping buffoonery (he of the split-pants fame), Tim knows more about honey bees than the average New York City bee-keeper. He also helped maintain beehives at a small farm in Red Hook called Added Value, and those bees, like David Selig's, were bringing in red goo and turning it into a metallic-tasting slush. Like David, Tim had gathered samples and sent them to the state for analysis.

All sorts of crazy stories about honey bees circulate around the world from time to time. Most at least begin from a kernel of truth. Way back a decade or so before the start of the twentieth century, there was a widely reported race between pigeons and honey bees to determine which could fly faster. As random as this may sound, during Victorian times, there were much stranger things happening in the British world. Women routinely bought packets of heroin gel from Harrods for personal use. The more fashionable among them wore hats adorned with stuffed cats and squirrels. Men raced against dogs in swimming matches across the Thames or wrestled bears (usually to the detriment of the former). Postmortem photography was all the rage, complete with propping up a dead relative, forcing his or her unseeing eyes open, and positioning old dead auntie or uncle among the living kinfolk to create a keepsake portrait. So the idea that beekeepers and pigeon fanciers might collaborate on such a project doesn't seem so outlandish.

Though the story originated slightly earlier in Europe, on April 16, 1892, *The Caulfield and Elsternwick Leader,* with offices in Melbourne, Australia, ran the headline BEES VS. PIGEON IN A FLYING MATCH. It described a race that took place in Germany (though it was northern England by some accounts and Belgium in others) where "almost a king's ransome [changed] hands" over the great debate as to which winged creature would home faster. The distance was sometimes reported as between two villages. The bees were supposedly rolled in flour first, so as to identify them upon return to the hive. Despite this obvious handicap, the bees, it is said, won handily, even though the two-winged fliers had been the favorite. "The first bee came in twenty-five seconds before the first pigeon. Three more before the second. The rest were not classed."

It was a great story that was reprinted for years all around the globe—and it was completely false. It was made out of whole cloth and swallowed by nearly everyone.

Back in Red Hook with the red honey, enter Cerise Mayo. Because *cerise* means cherry in French, some reporters initially had their guard up, understandably believing this tale to be a hoax. It wasn't; it was Cerise's hives on Governors Island that David Selig mentioned in his email to me. Gita Nandan, who along with Tim helped maintain the beehives at Added Value, sent the NYCBA some photographs of the red honey. The executive director of Added Value, Ian Marvy, also reached out. The beekeeping community in Red Hook was concerned not only for the bees, but also—to their credit—for the factory and the livelihood of the owners and workers as well. As Gerry Gomez Pearlberg, another Brooklyn beekeeper and poet chimed in, the situation was "definitely one of the more, er, colorful, beekeeping stories we've heard in a while!"

It was a fascinating problem, but it was not my problem. Until it was. In short order I heard from a man named John Bozek, who worked for the Business Outreach Center, a group in New York City that seeks to "improve the economic prospects of traditionally underserved groups, with a focus on low- and moderate-income entrepreneurs and their communities, and thereby create genuinely brighter futures." Though the cherry factory had been, at that point, owned by wealthy white guys for more than seventy years, most of their employees are minorities, so, maybe that is why Bozek and his group were involved. However John Bozek came to represent them, he wrote in part:

> We need somebody to act as a "bee consultant" in a difficult case in Brooklyn. One of our clients, a factory that has been in business for 50 years [sic] and employs 30 people in the community, has had a serious problem with a neighboring apiary.

So we set up a meeting with the factory owner. Since the media had broken the story, there were reporters camped out in front of the factory at nearly all times. To prevent a media melee, I asked if six A.M. would work. "I'll make it my business to be there," responded Arthur.

The day before the meeting with Arthur, I met with the Red Hook–area beekeepers to make sure we were all on the same page, and we were. They wanted to solve the problem for the sake of their bees, and they wished to be good neighbors as they gentrified the community in as sensitive and respectful a way as they could manage.

The first time I visited the cherry factory, Vivian Wang asked to accompany me. Vivian, who had begun beekeeping that year, was an attorney who worked for the Natural Resources Defense Council. In fact, she spent two years there, then left to work for the Department of Justice, and is now back again at NRDC. Born in Taiwan and raised in Texas, she moved to New York to attend Columbia undergraduate and law schools, and came armed with a master's in environmental policy from Oxford University thrown in for good measure. She is a champion of the environment and, in 2010, a newly minted beekeeper. So this was a fascinating problem for her and right up her alley.

We thought this was an unusual case. We were partially wrong. It turns out to be fairly common practice for the creative foraging of honey bees to result in honey that is unusual in both taste and appearance. In 1932, *The Brooklyn Daily Eagle* ran a story that read, in part, "Beekeeping in a very thickly settled city . . . would be difficult because there would be too much danger of the bees getting into mischief if they happened to pass a candy factory or any place where there was a sweet." Turns out it is true.

A World War II–era Coca-Cola factory in England had issues with honey bees foraging in their spillage and beekeepers ended up with a sickly grayish-brown honey-like goop. An M&M's factory in France enabled honey bees to create weird concoctions of oddly colored honey-like substances. A pallet of melted Popsicles in East New York resulted in green-hued honey in that area of Brooklyn. Antifreeze from a shabbily run auto mechanic's shop in Queens once produced blue-green honey-like poison. There's a long list of similar stories from all over the world, but still, the Red Hook story was compelling and stood out for reasons that will become clear.

Once again *The New York Times* blasted the lid off Pandora's bee-box. A front-page story about "red honey" from bees in the industrial area of Red Hook, Brooklyn, outlined how the honey bees were lapping up high-fructose corn syrup and Red Dye No. 40 from the runoff from a maraschino cherry factory, and, as a result, trouble was brewing for all parties. The bees' little translucent bodies glowed cardinal in the afternoon sun as they returned to their hives after a day of foraging, and the "honey" that they produced had an eerie hue. "When the sun is a bit down, they glow red in the evenings," Selig said. "They were slightly fluorescent. And it was beautiful."

My approach with Arthur was three-pronged. First, I worked with him at the factory to improve practices as related to insect access. We screened every possible opening to keep the bees out of the production area, and the factory stopped dumping the syrup into the street. Second, I ran interference with the fifty-plus media outlets seeking comment. Lastly, I was the intermediary with the beekeepers. The problem should have been cleared up as far as we could tell, but by then it was too cold for bees to fly to the factory to partake of any spillage anyway, and the damage had already been done to the beehives. The yield for the beekeepers was ruined for the season. Happily, the issue of red honey did not resurface for the beekeepers the following year, and the maraschino cherries continued to be shipped all over the world bee-free, without their cherry containers doubling as bee urns.

So all seemed well and good. Unfortunately for Arthur, though, his real troubles had just begun. Rumors, temporarily obscured by the red honey fiasco, began to recirculate that there might be marijuana being grown in the factory. One could occasionally catch a strong whiff of it in the neighborhood, but this was usually attributed to workers taking a toke on their break.

Arthur was a bit of an odd duck, too. His second wife had been a mail-order bride, and the woman about to be his third was a former adult film star. He was certainly a private person. Most people who have massive illegal marijuana grow centers in secret subterranean rooms have reason to be private. It turns out that the factory was equipped with a covert basement, not included on any plans or documents, "twenty-five hundred square feet, [with] space for about a hundred marijuana plants in a well-set-up system of hydroponic cultivation under L.E.D. grow lights," wrote Ian Frazier in *The New Yorker*. (In spite of my chagrin regarding Arthur's problems, I will always remember the flattering way in which Ian portrayed me in the article. "[Andrew] is a handsome, hazel-eyed man of French-Canadian parentage, with a suave black beard going gray." Free honey for life, Ian.)

When I worked with Arthur, he was anxious to solve the problem with the honey bees and acted quickly to do so. I appreciated his earnestness in that regard, whatever his real motivations. Sadly and surprisingly, when the authorities caught up with him, Arthur took his own life with a handgun in the cramped bathroom adjacent to his office while investigators were standing on the other side of the door. Now his two daughters run the company, the fourth generation of Mondellas to do so. Most who knew Arthur speak well of him. He seems to have lived high on the hog, shared what he had with those around him, and was certainly loved and is missed by many.

Normally by November, the beehives are wrapped up for the winter. The bees have slowed down from a busy year and begun to tighten their cluster, heating it to 95°F. Though they do not actually hibernate, they will come close to it. Beekeepers have reduced hives to make the cavities smaller so that the bees have less space to keep warm. We've long since removed the fall feeders atop the hive to avoid cold, dripping liquid landing on the cluster of bees. We've set up wind blocks to help the bees maintain warmth, and we've tightened crank straps in order to keep the hives upright and together, year round.

Tar paper is sometimes wrapped around the deeps, which are the larger boxes used with Langstroth hives. (They're larger because they serve as both brood and honey chambers.) The colony may be reduced down to one deep to conserve warmth, or two deeps if it's a larger colony needing more room to stretch out. In either case, a homasote board may be placed atop the top box to absorb moisture or maybe blankets or sawdust. The mouse guard is screwed into place to stop mice from entering, having babies, and eating up the wax. Some of us even say our prayers—because it can't hurt. We do all of this because if our bees die during the long winter, we may have no honey to harvest next year. We may have none to sell. We will virtually have to begin anew, perhaps with a package of bees from a farm somewhere down south, or split an overwintered colony from up north as the year begins again.

By November back in Quebec, *mon arrière-grand-père* Hector would have by this time placed wooden slabs on the dirt floor of his basement root cellar and carried down as many of his beehives as would fit. This would keep them safe from the wind and the excess cold so that they might better survive the long winter. Over the coming months, the family would enjoy freshly baked bread covered in jam and honey, slowing down and waiting patiently for the spring, just like the bees.

Far south of Canada in Manhattan, the previous spring I was hired to establish two beehives in the northwest section of New York's Bryant Park, close to where the Ping-Pong tables are set up. It's a beautiful park, and April through September, we hold monthly lunchtime talks for large groups that gather to learn about honey bees and related topics. By October the hives are put to bed. The park adds an ice-skating rink to the grounds late in the year, but one recent November, the temperatures were unseasonably warm and the bees were out flying. Of course, there was nothing available for them to gather, nectar or pollen wise. So they made beelines to the one place where they could find sweetness—the ice-skating rink.

Owen Harring, my liaison at Bryant Park, texted me: *Hey Andrew . . . we are having a problem with the bees and our ice rink. . . . Hopefully not another Maraschino cherry situation!* This time it wasn't cherries the bees were attracted to, but a coolant for the rink called glycol, which is sweet smelling, sweet tasting, and deadly to consume. In fact, there were many bees flittering over the ice, attracted by the scent but unable to penetrate the ice to reach the source of it. Owen was astute to discern that the source of their attraction was ethylene glycol. Lucky for us and the bees both, the weather became seasonally appropriate that week and the problem solved itself. Only a smattering of dead bees were found on the ice, probably the result of hit-and-runs with tourists' ice skates.

Most important, during my best November, my son, Nobuaki, wriggled his way out of his mama and into the world. I recognize he bears an unusual moniker. Nobuaki is written in two kanji: 允 and 章. 允 can be read Nobu (but is more commonly read *in*), but it's not really a common kanji these days. It means truly or truthfully; or to tolerate, to forgive. Some say this kanji was created from the shape of a human being captured and held with his arms at his back, in a situation where he is compelled to tell the truth. Other sources interpret it as the shape of a human with extraordinary talent, and that is how I choose to construe it.* There is a four-character idiom 允文允武, which means being truly versed in both literary and military

---

* All Japanese names have specific meanings depending on the choice and combination of the kanji.

arts, and this dual strength is something that I would like to see my cub realize.

Then we come to 章. It means to make things clear, distinction, and chapter (which makes ends and beginnings clear). To an extent, people come up with their own interpretation, but putting together those two characters, my translation would be "making the truth clear" or "truly distinctive."

Nobuaki was the given name of my martial arts master in Kyoto, Eguchi Nobuaki. It was a name which, out of respect, I never uttered aloud to Eguchi Sensei—I would never call him by his first name. Eguchi Sensei was a man who treated me like a son; I came to think of him as a second father. My Nobu was born five months after Sensei died. I thought it a meaningful way to keep the name and spirit alive. The first Nobuaki grew up in the ashes of the Second World War, knew great personal loss and hunger, and was a disciplined and learned man. He also had a fantastic sense of humor and devotion to his community. He was a man worthy of respect and love. So it is a felicitous name for a much-wanted child.

A few addlepated plebeians criticized me for not naming my son after my own blood father. Some people voiced their objections to a little white boy having a Japanese name. Aside from having lived nearly a half century not necessarily adhering to other people's values, I had discussed this topic with my father. In his usual pragmatic fashion, he quickly concluded, "Let the name 'Norman' die with me. I hate my name."

It is my hope that young Nobu will follow in my footsteps and learn to work with and love and respect the honey bees. At least I would like for him to fully absorb the hard work ethic for which honey bees are known and celebrated, and which beekepers must embrace if they are to have any degree of success. Nobu has already been with me on several beekeeping adventures. There is a wonderful photograph of him shaking hands, at eight months old, with the president of the UN General Assembly. Nobu met His Excellency Peter Thomson when we ironed out details to place an apiary on the grounds of the international territory of the United Nations. During that meeting, Nobu, who had recently learned to walk and indeed trot, did so among the flags and furniture of His Excellency's office while his attentive and sharply dressed staff made sure that Nobu didn't collide with sharp corners. A fifth generation Fijian by birth, Peter is a down-to-earth, warm, intelligent, and hardworking man who, in addition to many accomplishments prior to his appointment to the United Nations in 2010, served in Tokyo at the Fijian embassy for four years in the 1980s.

The idea to place the beehives was the brainchild of his wonderful wife, Marijcke, who worked with me from the start, throughout the installation, and beyond. She was aware of Bees Without Borders after having read about a program my father and I had initiated in Fiji several years prior. The Fiji BWB trip was the first time that we had traveled to a place free from war or conflict to do beekeeping classes and workshops, and it proved to be a wonderful experience. We worked on the main island of Viti Levu and in the Yasawa Islands, including Turtle Island. We fell in love with Fiji and Fijians, while spending time pulling feral honey bees out of trees, managing neglected colonies, and teaching beekeepers a bit about marketing their honey—not in upcycled plastic water bottles—to the honeymooning tourists who would, and did, pay top dollar for it, providing much-needed surplus income to the Fijian beekeepers.

Anyway, while planning the UN apiary, Marijcke and I met at their apartment and in meeting rooms across the street from the UN to strategize. Of course, I also had to sit in many meetings at the UN, mostly with the companionship of Katherine Morris, former army, the wife of Australia's deputy defense attaché, and one of NYCBA's beekeeping apprentices. We went from office to office, long meeting to longer meeting, month after month, and finally the United Nations, an organization known for seemingly insurmountable red tape, gave the green light.

For the actual installation of the beehives, in addition to Katherine and my father, several of my beekeeping buddies lined up to assist. It was no small joy to be on the North Lawn of the United Nations, an area not open to the public. The beehives themselves, I felt, needed to be painted the blue for which the United Nations is known. I called on my young friend Tristan Pinto, brother to Sasha (who had assisted me at MoMA in creating those small building-like beehives) and a creative fellow in his own right. Tristan came up with an eye-catching three-dimensional design that showcased the Bees Without Borders logo on the fronts and backs of the beehives, and beautified the trinity of boxes in a way that I could not have. He also helped with the bee work at the opening ceremony, as did Sasha and Hannah, of course Allison, and, naturally, Norm and Nobu. Along with Peter Thomson; his wife, Marijcke; and a large group of others, we celebrated the installation of extraterritorial honey bees as they represented many of the sustainable development goals of the United Nations.

The bees are not the only residents of the North Lawn. In 2016 an eyesore of a building was removed to open up the expanse of green again, making room for a number of sculptures and statues from around the world. The first was *Arrival*, a bronze ship cast in depiction of a potato famine–era coffin ship. The 2008 creation of Irish artist John Behan, it was the sort of ship that my maternal grandfather's family would have taken to immigrate to the United States from Ireland when the famine forced them to flee or die. Seeing it a stone's throw from the apiary always reminds me of a family of Irish beekeepers, notably Micheál and Aoife Mac Giolla Coda, who are well-known breeders of native Irish black bees in the Galtee Vee Valley.

Whether due to his love of dragons or as a result of his maternal lineage, a favorite of Nobu's among the works of art on the North Lawn is the Soviet Union's 1990 contribution of their interpretation of Saint George fighting a dragon. In this piece, the dragon represents nuclear arms, and "is made of decommissioned Soviet SS-20 and US Pershing II missiles. It's kind of like DIY, only with dangerous military waste," according to BuzzFeed News.

In November 1989, I was a teenager backpacking through Europe when the Berlin Wall came down and citizens began dismantling it with hammers, shovels, and their bare hands. I still have a piece of it from then. Fast-forward three decades, and a huge section of the wall now resides on the North Lawn, between a statue of a fellow on a horse from the former Yugoslavia and an avatar-like figure from Mongolia. I take no credit for the Berlin Wall being demolished, other than the one small piece that I wrested from it, thanks to those who enthusiastically demolished the barricade. However, twenty-eight years later, while reversing my truck on the pathway while servicing the UN beehives, I nearly took a section of that iconic wall down for the second time. I had some wood sticking out of the back of my pickup truck that brought me a bit too close for comfort to initiating an international incident.

Some at the United Nations would have been appreciative had I reversed into the one truly scandalous sculpture on the grounds. Just west of the apiary is an elephant statue with an impressively engorged penis the size of a thick tree trunk. Honestly, the elephant appears to have five legs, or four plus an enormous kickstand. At first I thought there must have been some extremely attractive female elephants around when the thing was sculpted, but the truth is even stranger: A live elephant was drugged and cast to make the sculpture. In 1998, somehow Nepal, Kenya, and Namibia got together and decided to find an African bull elephant, put it to sleep, and make a cast of it. The process, for whatever reason, stimulated the libido of the Proboscidea. The result is a statue near the beehives that shows the proud elephant with a full-mast erection. I wish I had been there to hear the discussion as to how the situation would be handled, but in the end, the powers that be at the United Nations found a decidedly diplomatic solution. There is a generous amount of shrubbery strategically placed around the supremely well-endowed statue, and modesty prevails.

There's some irony in the apiary being placed in such close proximity to that lusty pachyderm: Elephants and honey bees do not get along. In fact, honey bee cultivation has been used in Kenya and elsewhere as an elegant solution to a dangerous problem of elephants trampling villages and killing people. This has become an unexpected issue because villages and homes have been built in the elephants' traditional migratory paths. To avoid inadvertent trampling of people, ruination of crops, and conflict between people and elephants that leads to the harm of one or the other or both, fences laced with beehives have been put in place near crops and homes in several countries where elephants journey. Sometimes there is a rope attached to the beehive, and that is stretched across an area near crops. If and when a hungry wandering elephant jostles the rope, the colony is disturbed, and the bees repel their big-eared foe. It is both a simple solution and an effective one. I've worked constructing some of these beehive fences myself. Elephants approach, encounter the bees, and turn in another direction without any real harm to any party. Elephants are hulking creatures, but they do not enjoy stings inside their trunks or ears.

So I like to think of the beehives at the United Nations as serving multiple purposes. They draw attention to the Sustainable Development Goals of the United Nations. Their residents are the reason for World Bee Day (May 20!). They pollinate the rose garden and provide honey for the diplomatic corps of the world. And, finally, they keep the diplomats safe from a certain randy bull elephant that might trample or otherwise interfere with their peacemaking.

DECEMBER

"The only reason for making a buzzing-noise that *I* know of is because you're a bee."

Then he thought a long time and said: "And the only reason for being a bee that I know of is making honey. . . . And the only reason for making honey is so as *I* can eat it."

—A. A. MILNE, *Winnie-the-Pooh*

Generally being a law-abiding sort of guy, I was surprised to find myself in the hot seat in a small, windowless interrogation room in Harlem's Twenty-eighth Precinct one morning in 2018. There was a white table with links bolted to the top to secure handcuffed suspects, and a small one-way mirror recessed in the grungy wall. While I was not handcuffed, the door was closed and I felt slightly claustrophobic. I sat waiting and rereading an English translation of "In a Grove," a short story by Ryūnosuke Akutagawa, from a book that happened to have been in my back pocket and wasn't deemed a weapon (though knowledge is power). In the story, a man is detained by the police and is interrogated. The larger theme is how one event is seen differently through different sets of eyes. In this sense the character's situation and mine were similar, though mine lacked the swordplay, rape, and murder.

I had been one block away from the school my son attends on the Upper West Side near Central Park. I was in the company of my friend Jim Pletcher, a recently retired associate provost at Denison University who had been apprenticing all year with the NYCBA. Jim and I had been standing next to his car, which was parked near mine, moving beekeeping equipment from one vehicle to the other. I spotted three men walking briskly toward us.

I quickly surmised that they were among New York's finest. They may have been in plainclothes, but it was obvious from their comportment, if not the badges clipped to the belts or bouncing from chains around their necks, and from their clearly visible firearms, that they were the law. They were all a bit younger than me, fit, and unshaven. They were pretty friendly, in fact. But, as Jim later said, "They looked like they could be pretty unfriendly, too." They had been sitting—for who knows how long?—down the block from my pickup truck waiting for me to arrive. With my plans for the day now altered, Jim and I shook hands and parted ways, he in his car, and me and my three new friends in their vehicle, at their insistence.

After establishing that I was Andrew Coté and not some other short, pudgy, gray-haired bearded beekeeper hauling beekeeping equipment around Manhattan, they said, "We just want to ask you a few questions at the station," as if I was in a *Law & Order* episode. Two of them bookended me and hustled me toward a soccer mom–style minivan they were using that day. The third hopped into the driver's seat.

If they wanted to be undercover, that ride was the right choice; it certainly was a departure from what I would consider a law enforcement vehicle. If I saw that thing pull to the side of the road, I would have expected a few kids with sports equipment to hop out of the side before I would ever have imagined three armed detectives emerging from it. But whatever was happening, it was surely serious, given that the NYPD had dedicated three men to sit and wait for me until I showed up to my truck, which had been parked in that spot for a few days. I decided against using the "What's this all about?" line. I did not technically have to go with them, since I was not under arrest. But I thought that it was best to just clear up whatever problem there was. They might have found me at a less opportune time—when I was selling at the market, for instance, or when I was visiting a client, or with my kid at the playground. Better to get it over with, I figured, whatever it was.

The detectives were friendly and chatty as they drove me to the station, and since I did not know why I was there, I kept my responses vague. They informed me they were not the primaries on the investigation for which I was picked up. They didn't seem to mind indirectness any more than I fretted over their vagueness—they weren't interviewing me—but they did have a lot of questions about bees.

This is a fascinating constant for me. On occasion it borders on being a problem. On the one hand, I am delighted that people are interested in honey bees, and what I do, and how the whole enthralling and sweet world of beekeeping works. People are especially interested in urban beekeeping and have no end of questions about it. I want to always be positive and endeavor to take the questions and the person seriously, and be grateful for their interest.

Still, the truth is, when I can get away with it, I often avoid telling people what I do for a living. I skip dinner parties. I even try to avoid telling a doctor during an appointment or a barber during a haircut what I do, because I fear that I will be, as I usually am, met with a barrage of questions that are sometimes too much for me to answer. As fascinating as honey bees genuinely are to me, at a social function I might some-times want to talk about a film, an exhibit at the Museum of Modern Art (unrelated to bees), politics, or anything else. I don't wish to sound ungrateful—I know how fortunate I am to do something I love—and it is very nice that people are inter-ested in it enough to, say, buy a book about it—but sometimes I need a bee break. In fact, one of the reasons that Yuliana and I get on so well, I think, is because she is not interested in the ancient art of the apiarist, or "Whatever it is exactly you do all day on the rooftops with your—your bees," as she puts it.

An exception to this is when I want to keep law enforce-ment in a good mood. Back in the unmarked minivan that one of the detectives had, it appeared, borrowed from his mother, I answered bee questions with a big smile. At their promptings I explained how the bees go about gathering the nectar to turn it into honey; just what propolis is; and what bees do during the winter. (This is something that many, many people wonder and ask about.) Then I started to wonder about my own winter. I was due the next day to go down to an island in the Caribbean to establish an apiary on a private farm on Cayman Brac in the Cayman Islands for a Serbian businesswoman. She had established a young orchard and wanted to ensure pollination. And she wanted local honey. There were no honey bees on that island, and the main beekeeper on Grand Cayman refused to sell her any. The plan was to bring a couple of dozen beehives via boat from Cuba right down to the smaller island of a country best known for its shell corporations and being a tax haven. Now, instead of the Cayman Islands, I feared ending up on Rikers Island, New York City's notorious jail where 85 percent of the daily average of 10,000 inmates have not yet been convicted of a crime. I didn't want to make it 10,001. I wanted to drink Cuba libres and remain free myself.

Sometimes the Department of Health and Mental Hygiene seeks me out to assist with honey bee matters. Their inspectors seem to be all good people but have no training in and no extensive knowledge of practical beekeeping, and no budget is allocated to change that. Among other matters, several times per year the department asks the NYCBA, or me specifically, to assist in removing beehives. If indeed it is a colony of honey bees and not a wasp nest, it is usually the same old story: Someone has placed a beehive on property that is not theirs. The property owner, not knowing whom to contact, tries various city entities. Eventually the property owner is directed to DOHMH. DOHMH checks its books and tries to locate the owner of the beehive. If the hive is not registered, there is no way to know to whom the beehive belongs. DOHMH then asks the NYCBA if we know to whom the colony belongs. We check our records and reach out via social media, giving not quite enough data to identify the location to an opportunist, but enough clues that the true owner would recognize it; this to avoid someone claiming a hive that is not theirs. Sometimes we put a note on the hive and wait a week or two. Our approach depends on the situation. If all of the above fail, DOHMH asks us to physically take possession of the beehive and remove it from the premises. "This falls somewhere under the purposely loosely worded nuisance clauses in the regulation," said Nancy Clark, who was the primary force within DOHMH that most helped us to win legalization. Nancy is now retired and has gotten big into the grandmothering business. She stops by my honey stand at Union Square regularly, and we swap bee and baby stories.

Anyhoo, one time DOHMH told us that a particularly negligent person had placed two beehives on a fire escape. Those we removed within three hours without any attempt to find the owners. We all felt (DOHMH included, of course, as it is always their call) that a citizen imprudent enough to place a beehive on a fire escape should not be handling bees. That person, in fact, probably should not be allowed to mix with the general public at all without supervision. Someone who is fleeing an apartment due to smoke and fire has enough to worry about without also, in blinding thick smoke, tripping over and upsetting a colony of fifty thousand angry bees. So in those sorts of situations, at the direction of the DOHMH, we just take the hive without any attempt to enlighten the former owner. Then there are beehives placed on rooftops of buildings with clear signage that warns to stay off the roof. Or placed in community gardens without permission. Or on private land. There is no set formula to the rules of removal used by the DOHMH. The only commonality is that sometimes people act in ways that defy common sense—which itself is remarkably uncommon—and just set up shop with their beehives wherever they find a space, whether it's legal or responsible or not.

And what becomes of those beehives? It depends. They have to be put somewhere. Inheriting a beehive in this manner sometimes is a windfall, but always, it is a responsibility and a potential liability. Whatever diseases the colony may have will probably be passed along to whatever other colonies one has nearby. The unknown hive may be aggressive and interfere with other colonies or the public. Depending on the time of the year, the value fluctuates greatly. If the bees are about to start foraging, it could be a big win for the receiving beekeeper. If they are about to settle in for the winter, the inheritance is not so valuable. There is something to be said for equipment that is well cared for and in good condition. Usually, though, in my experience, the beehives that are placed by people who do not have the good sense to get permission from the property owner are not in stellar shape. Generally when we appropriate these colonies the bees themselves are rehoused and the equipment is discarded or donated to a community garden or to whichever beekeeper is at hand and willing to haul it away—that is, if the entirety is not donated.

One such case where DOHMH requested assistance was in Harlem in what appeared to be an abandoned lot. Later, it was suggested that it was a registered community garden, but it displayed no plaque, no signage of any kind, and bore no resemblance to anything other than a pile of dirt and refuse in a space between two buildings on a dilapidated block. DOHMH did not consider the ground more than a vacant lot.

So there I sat, in the stagnant little room within the detective squad on the second floor, alternately reading about how Tajōmaru confesses to some of the crimes of which he is accused, and nodding off. After about an hour, Detective Lonnie Brown entered the cramped room and took up about half of it. Detective Brown is a tall, handsome, powerfully built African American man, and he seemed like a swell guy. I know plenty of police officers and have been around them all of my life. My father's brother Irv, who is also my godfather, was a police officer. My brother, Mike, is, too, as are many of his friends, of course. I have known loads of them, so I feel comfortable around them. I did not, however, mistake my comfort as a guarantee of their good nature or their investment in my best interest. Still, I knew I hadn't done anything wrong. Whether it's fair or not (it's not), I knew I had relatively little to fear from Johnny Law given that I was an innocent, middle-class, middle-aged white man. I really felt like Detective Brown was simply trying to figure out a few things. So he sat down, and we started to chat. But I wasn't quite ready to throw caution to the wind and spill my guts. His easy manner and empathetic nature were possibly just part of his interrogation style.

"Do you know anything about some beehives that were taken from a spot on 119th Street?" he asked directly. The truth is that I did know about it. All about it. Months earlier I had received an email from one of my several contacts at the Department of Health, who sent me several pictures and wrote:

> Good afternoon Andrew,
>
> One of our staff members came across these aban-doned active hives (there are live bees) . . . located at 13 West 119 Street, Manhattan. . . . Please let me know if and when you can send someone at the loca-tion to investigate and remove the hives.

I missed the first email but they re-sent it a week later, at which point I had followed up and made sure, in writing, that the DOHMH bosses knew about the request to remove the hives. They did. So I went there myself with a few of NYCBA's apprentices. Sure enough, we found three live beehives in deplorable condition. I checked the insides of the boxes and the tops of the frames for a name—sometimes beekeepers write or burn names or initials into the equipment—but this is more of a commercial beekeeper's practice than a hobbyist's. Although apparently orphans, the bees seemed healthy. How-ever, the wooden boxes were tilted back, allowing in moisture, which causes a whole host of problems, including rotting wood. They were on or near the ground, which was also con-tributing to the rotting wood, and they seemed fit to burst with activity, so they had either overwintered well or were recently installed packages. In either case, they needed room to grow or else they would soon swarm. Indeed, it was later discovered that one colony had swarm cells already.

I wanted to put a note on the gate of the seemingly abandoned apiary and give the owner(s) a chance to relocate them, if, in fact, there were still owners. But DOHMH responded negatively to my request to do so. They wanted the bees removed immediately. I never asked them why, but I assumed it had to do with the fact that they were only a few inches from the property line to the neighboring lot, where there was construction taking place on what had clearly long been an abandoned building. I was later told by the NYPD that the hives were considered abandoned property, by the police department at least; DOHMH considered them abandoned simply by virtue of having no registered owner. Whatever the reasons behind their desire, the authorities wished them moved posthaste.

It was springtime when all this happened, so I was slammed with work. Rather than handle the removal myself, I contacted three of my fellow beekeepers. They appeared at the site at daybreak one day, crank-strapped the hives together, screened the bees in, and started to carry them the thirty or so feet to a waiting pickup truck parked halfway on the sidewalk. It was child's play to do this, but as my father always tells me, "There are no five-minute jobs." The morning calm of the two men carrying that hive vanished as both the figurative and literal bottom fell out and thousands of angry bees let their displeasure at the disturbance be known. One of the beehives, screened and strapped securely, was so thoroughly rotted out on the bottom that when it was lifted and moved a few steps, all present found themselves under full assault. Perhaps the former owners can find some emotional compensation in the fact that a couple of the guys ended up with swollen ankles that day from a plethora of stings given by their former bees. It is all the reparation they will ever receive.

To my knowledge all of the equipment was discarded, primarily because it was pine (a softwood) and left exposed to the elements at least all winter. There were additional boxes, too, which had been left out in the elements and not even covered by a tarp. It all amounted to a mildewy, dry-rotted, undesirable stack of wood. But the bees themselves, unaware of their impoverished state, were healthy and happy and were put to work immediately in new and certainly improved lodgings. When a car is stolen, it is sometimes taken to a chop shop and broken up into pieces, and ends up being worth more in parts than it was as a whole. This is not the case with appropriated live colonies of honey bees. Only the bees and the frames they were on remained; the rest was tossed. Then spring continued.

Upon questioning, I told Detective Brown that I had heard from DOHMH that the beehives needed to be removed, and I had passed that information around the NYCBA, and several people had been interested in relocating the beehives. I told him that I had been to the spot myself and had seen the apiary, which had appeared to me also to be abandoned. In any case, the beehives had been moved to three different apiaries, in two states, by the end of the day.

Months after the hives were moved, I was standing at my booth at the Union Square Greenmarket when a fellow I'd known peripherally for perhaps eight years wandered up to my tidy little stall. He was a regular at the bee club meetings and often purchased packages of bees through the bee club or from me. He was a nice enough guy, and he and his partner lived in my neighborhood in a nice duplex. I sometimes saw them walking their two mastiffs in and around Central Park. When I asked him how his bees were doing, the floodgates opened. He told me a whole long saga about how his hives had been stolen.

I had no idea when he started talking that I had helped arrange for his bees to be taken; I am called upon so often for similar jobs that once they're in the past, they often blend together. When I eventually started to get an inkling that I may have been involved, I asked him some specific questions to try to ascertain whether there was a connection. He told me that he had filed a police report, but he didn't know which precinct it was he'd visited or even where it was. That seemed strange to me. He said that his hives had been registered, but I knew that no hives had ever been registered at the address I'd visited in Harlem. He told me that the property where his hives were located was an established community garden, but at the time there was zero signage at the Harlem location I'd visited. A call later to GreenThumb, which oversees all community gardens in the city of New York, said there was a pending application for the Harlem site, but there were problems with it. So I figured this must be a different location, a different situation.*

"Too bad," I thought, sincerely. Not a unique tale. I've had

---

* As of 2019, the site has been cleaned up, the ground leveled, a gazebo installed, a slate pathway added, flowers and tomatoes planted, and a GreenThumb sign affixed to the metal gate, declaring the former junkyard apiary as the Walter Miller III Memorial Garden (La Casa Frela).

beehives stolen from my farm in Connecticut, and the theft of hundreds of beehives at a time from rural areas is a sadly common occurrence. Some beekeepers now outfit their beehives with GPS or other tracking devices in case of such larceny, but not most that I know, and not people with only a few hives.

So Detective Brown and I went through the entire episode, and he excused himself. He asked me if I wanted a drink. I declined but thanked him. Sometimes suspects are offered a drink to fill up their bladders, which puts them ill at ease. To further unsettle them, they might not even be permitted to use the facilities in a timely manner. But I'm guessing that was my paranoia more than anything else. Probably Detective Brown just thought I might be thirsty.

So I sat in the interrogation room for a good while longer wondering what would happen next. One aspect of all of this that bothered me was that I was supposed to meet up with my father later that morning and, assuming I could reach him, I didn't want him to worry about me sitting in the Harlem police station. He had many similar experiences with the police, with no out like I had. Dirt poor, as a teenager he and his best friend, Eddie—my mother's brother—would often walk at night. It gave them something to do, and they might walk and talk for hours. My father is slightly bow-legged and Eddie is somewhat knock-kneed, so the family joked that walking side by side they spelled the word "OX." Sometimes the police would pick them up and take them into interrogation rooms just to mess with them. Anyway, I didn't want him to worry needlessly.

Another hour later and again true to the script of every television crime drama that has ever been aired, Detective Brown reentered the room and said, "Okay, you're free to go." I was much relieved but not surprised. "I just got a call from DOHMH. They confirmed they authorized the removal." I put my book back into my pocket as I stood up. "I guess you need to get back to your market," Detective Brown said, smiling, showing me that he knew my schedule. He walked me out and we chatted about bees some more, and I answered his questions, which had been informed by videos and articles he had read about urban beekeeping—and me—as he pursued this case. He confided that he hadn't bothered to work the case sooner because it seemed a bit ridiculous to him, and once he started investigating me, he realized that I would probably not "risk it all over a few beehives." My gut feeling that he was a good guy was correct. The beekeeping neighbor who accused me of stealing his beehives was still not at all happy when I ran into him shortly thereafter and told him what had transpired. I do not know if he was upset that I was not charged or that he had lost his apiary. Or both. I did find out that it was his beekeeping partner who had filed the police report, so he had come to me at the market sort of half-cocked and under-informed. Still, I figured we might be able to help him out as perhaps the next few wayward beehives, or at least swarms, could be passed to him to help ease the sting of having lost his own.

December is usually filled with all sorts of activities related to holidays and end-of-the-year celebrations. Usually this results in a rush of shopping, which means that the Union Square Greenmarket can remain quite busy throughout the month. One evening, despite business being as brisk as the weather, I needed to close my stand a little early because I was scheduled to attend a function at the United Nations. The previous year I had installed the apiary on UN grounds, and since then had replaced the center Langstroth beehive with a traditional Slovenian double beehive. Slovenians are serious beekeepers with a long tradition of maintaining colonies and harvesting fine honey. When they decided to introduce the idea of a worldwide day of recognition of the humble honey bee and her cousins, they donated a gorgeous traditional Slovenian beehive to Bees Without Borders, which I placed in the apiary at the United Nations. There was even a special honey bee stamp issued by the UN post office to commemorate the occasion and celebrate the honey bee.

I was very happy to have been invited to the gala at the United Nations celebrating the passage of the proposal for World Bee Day. Though it was December, and May 20—the agreed-upon date of the celebratory day—was far off, a vote had been cast earlier that week, and some delegates from Slovenia who were themselves beekeepers had made the trip from Europe to celebrate the event. It was nice because we all knew one another by reputation, and when we met for the first time it was like meeting old friends. The camaraderie found among beekeepers is refreshing and encouraging in a world that is often divisive.

That day was the last Wednesday before Christmas, so many people at the Union Square Greenmarket were buying last-minute gifts, and I was happy for the business. I'd carried a suit with me to the market, but forgotten to bring appropriate dress socks. When the sun went down and I was done at the market for the day, I climbed into the back of my Honey Mobile—a converted U-Haul onto which we'd hand-painted scenes of honey bees buzzing high above Manhattan as a sort of rolling, roving billboard—pulled the rolling gate shut, and donned my suit. My one nice pair of Brooks Brothers Italian shoes just barely fit over my thick gray wool winter socks. I emerged from the rear of the vehicle a new man, confusing the few people who happened to be standing by my truck.

I drove crosstown and left my worker, Zeke, to sit in the truck for three hours since finding legal parking for an oversized vehicle anywhere near the United Nations is nearly impossible. So Zeke generated a little extra income while napping and babysitting my truck, and I drank red wine and hobnobbed amid a couple hundred finely dressed people, including a few Balkan diplomats and a motley crew of local beekeepers whom I had invited to join me at the suggestion of the hosts. We all cleaned up reasonably well. Possibly we lent an air of rustic authenticity to the festivities, but more likely we weren't noticed by the diplomatic crew.

Among my guests was a guy named Flynn, who worked in security at the United Nations. Flynn was raised by a single mother who was a police officer in the south Bronx during the worst of the crack epidemic. He skated through Desert Storm as a marine without incident, and then survived being shot in the Bronx while he raised three daughters as a single father. He still lives in the Bronx with his youngest, Alaura, who was there to enjoy the evening with us, too. In their apartment they keep beehives in a spare bedroom. The bees fly in and out of the windows and do quite well for themselves, feasting at Van Cortlandt Park and Woodlawn Cemetery. I am unaware of anyone else who has placed an NYC apiary indoors with sustained success. I do not mean a simple observation hive with a hose out a crack in the window—I mean full-blown hives in a screened-off room. It is a remarkable testimony to the intrinsic neurosis of apiculturists. And supports Alaura's assertion that her father is nuts.

Then there was Toby Bloch, an iconoclastic country boy pulled to the big city by his hopeless love for a Prada-wearing better half, Daniela, and seeking to reconnect with his home-steading roots by keeping bees on his roof with his daughter, Olivia. He loves explaining to anyone who will listen how bees epitomize the concept of collective action, maintaining a net-work of mutual care. His favorite pastimes are sharing what bees have to teach us about a socialist utopia, and trying to instill a sense of collaboration for the greater good around issues of housing, transportation, and environmental justice. So in essence his values align with those of the honey bees perfectly.

My beekeeping friend Robert Deschak attended and was even kind and generous enough to provide me with a lovely necktie adorned with honey bees embroidered in gold thread, which I, of course, changed into immediately. Like bees in a winter hive we huddled together much of the time, but more to chat than to share warmth and food, though we did that, too.

Yuliana came along to impart an air of class and respect-ability to the entourage. Like me, she had left work a little early, then walked in the biting cold to meet me in front of the

Japan Society, half a block from the UN. We then met the rest of the retinue in front of the 193 flags that adorn the perimeter of the international territory in the Turtle Bay area of Manhattan. We all went inside and through security, which resembled TSA at an airport, complete with bag X-rays and metal detectors. We entered the area designated for the party, grabbed some wine and fancy hors d'oeuvres, and were immediately impressed by the wonderful honeycomb sculpture that had been erected for the event. It was massive, large enough for dozens of people to stand in. More than that, it was a learning station where people could see videos, play interactive games, and even wear virtual reality goggles that would transport them into a beehive, as it were.

At one point Yuliana and I were inside the honeycomb sculpture when I sensed a malevolent rumbling from below. I glanced downward to find Liane Newton. Liane had come onto the beekeeping scene around the time legalization passed, though she was not involved in that effort. Appearing around sixty years old and an attorney who as far as I could find has never practiced law, Liane was living with her mother on the Upper East Side. She is the current head of another beekeeping group, which has a name that was deliberately created to be confusingly similar to the New York City Beekeepers Association. She came over to say hello. Sort of.

Leashed in pearls, she sashayed over to us with all of the grace of a hobbled penguin. "Andrew, why, I hardly know you," she began, as I craned my neck downward to hear her. "You've grown a beard and you've put on a *lot* of weight." I had not expected to be fat-shamed on international territory so early into the evening. But I could not deny the truth of her observation, either. In respect to the latter, in Japan my new bulk is called *shiawase-butori,* or happily plump (幸せ太り). The delightful and apt expression refers to the weight a man gains that is the result of someone new in his life who's feeding him and making him happy, healthy, and comfortable enough to let himself go a little. I must have been very happy at that point—euphoric—if we were to judge by the snugness of my suit. The irony was that this commentary came from a woman who, as a bulky bantam beekeeper, commands greater heights supine.

When two queen bees are thrust together due to whatever circumstances, usually they will fight to the death. On a rare occasion, however, one queen, rather than fight, will hoist her abdomen and point it at the opposing queen. Then, in an act called spraying, she will hose down her opponent in fecal matter. I understood for the first time how that poor queen on the receiving end of that feculent flow must feel.

Even though we'd had only one prior conversation, which had taken place seven years before and had lasted not even fifteen seconds, Liane had been a mild nuisance over the years. A large component of the irritation factor was the name of her group, though there were certainly other factors. She and her partner, Poindexter Pantstootight, he from back at the Fort Greene broken branch fiasco, decided to name their beekeeping group after ours, changing only three letters. She and Poindexter had some sort of relationship that came to a bitter end, and she'd ended up carrying the torch for their once-joint club. After a mercifully brief exchange, Liane descended the sculpture and buffaloed into the crowd to spray her charms elsewhere.

I breathed deeply through my nostrils and filled my chest. "How in the world . . ." I paused and regrouped my thoughts. "How did Liane come to find herself invited?" I asked my contact at the Slovenian consulate through a clenched smile. Yuliana was holding my arm and gently dug her nails into it.

"Oh, she isn't with you? She's not from the same group?" was the genuine response. The misappropriated club name had done me in again! I threw back more wine and smiled. Yuliana offered, "It's always nice to run into old friends," and we both laughed.

That short encounter aside, the UN event was a lot of fun. The fact that we now had World Bee Day (May 20), an internationally recognized day to celebrate our beloved honey bee (though the day is for all *Apis* species—National Honey Bee Day is August 17), was certainly a cause for a grand party. As I gazed at the diverse group of people in attendance, including the eclectic group of beekeepers who had come with me and who were enjoying themselves so much, I felt that perhaps leaving academia and jumping headfirst into a world of four-winged, five-eyed insects may not have been the worst idea I'd ever had. I tried to harbor a glimmer of hope that in the coming years the New York City beekeeping community could unite and enjoy the sweetness of our shared interests more than we've been able to over the past decade.

But most of all, I saw that moment as an appropriate launching pad for the growth of Bees Without Borders. What better crowd to facilitate furthering our work with the international beekeeping community than a group of international beekeepers at the United Nations? I had never harbored aspirations to grow BWB, as I was happy with my lifestyle such as it was, but there seemed to be a higher calling or purpose beckoning, perhaps. Or maybe I had drunk too much wine. But requests for workshops and offers for collaboration were piling in.

As December draws to a close I remember one of the more simultaneously productive and problematic BWB trips undertaken in the twenty-plus years since the organization's inception. My father and I were in Samburuland, in Kenya, staying in rudimentary lodgings in a remote area and cooking our own meals. One day I purchased a live scrawny chicken, whose feathers weighed nearly more than its meat, as it turned out. That evening I sliced off its head, and we cooked the poor bird for dinner. We pump-filtered all our water—for drinking, cooking, and brushing our teeth—on the grounds of an orphanage where every one of the children has some sort of heart trouble. The scene was bleak. They lacked adequate food, clothing, and medical supplies, and for play they kicked around a sort of homemade ball made from who knows what wrapped in tape. The nonspherical object didn't roll; it just klunked over a few times before coming to rest on one of its uneven sides.

"At least it won't roll into the street," Norm quipped. "Not that there are too many cars." (There could be days between cars.)

But of course it was Norm, who grew up hungry and poor, who remembered this pathetic little ball-esque thing in particular. So on our second trip there he brought along deflated balls of all types and a hand pump. I'd rarely seen him look happier than he did sitting in the speckled shade of an acacia tree, choked with children surrounding him, slowly inflating one ball after another and handing them out and tossing them around. We had brought suitcases full of donations for these kids—clothes and shoes, mostly—many of which had been donated by students and faculty from the high school I had dropped out of, Brien McMahon. But nothing was as well received or enjoyed as much as those balls.

During our second trip to Samburuland, we worked with the Samburu and dug deep into the ground to install wooden fence posts and a heavy wire fence to keep the honey badgers at bay. The previous time we had been there, shadowed by a CNN film crew, we distributed beekeeping equipment such as veils and gloves and beehives, and installed them, and they had been doing very well. Before too long, however, the dreaded honey badger had found the hives and tore their metal lids right off and pounced on the larvae within. Fully half of the beehives had been destroyed by the time we returned. To protect the investment of the beehives, we set to work building a huge fence around this and two other apiaries. The first order of business was to dig a trench three feet deep all around the perimeter to bury the wire part of the barrier to keep the honey badgers from burrowing under the fence. It was thirsty work, but many hands made it light.

At night we had nothing to do but talk and watch the stars. There was no Internet or television, of course. There wasn't even clean water without real effort. On the evening of December 31, Yuliana suggested that we all go outside and look at the blazing stars. We did, and thus the old year passed into a new one just like that, our necks stretched up, the stars so bright, time marching forward with nothing and no one who could stop it or even slow it down.

Though I was in sub-Saharan Africa that night with two of the people who matter the most to me—Yuliana and my father—it was not all joyful. While on this trip we received the sad news that Aldea, my father's mother, had passed away at nearly a century old. We weren't prepared to lose her but she was ready to go. A doctor had recently told her that she would easily live to be a hundred. "I'd better not!" she responded fiercely. The story made us laugh, but she was not joking. She'd had enough. She was later cremated and buried in the low-lying spot in the Norwalk cemetery right beside the same train tracks where her sister Aline had died and been buried in 1936. Eighty-plus years after their separation, the sisters were together again.

In several ancient cultures, honey bees are believed to be a conduit between this world and the next. It is an old tradition among beekeepers, since at least medieval times, and particularly from Ireland to Germany, that when a beekeeper shuffles off this mortal coil, the bees must be informed. In fact, all great life events—marriages, births, deaths—were related to the bees. The hives were approached and knocked upon sharply with the knuckles, and the bees informed verbally of the demise of their keeper. Failure to properly put the bees into mourning—which aside from the telling, often included leaving some of the funeral bread in front of the hive, or even draping the hives in a cloth—would supposedly result in a poor harvest, a swarm, or even the entire colony absconding. Though Aldea had no more beehives of her own at the hour of her death, I mentioned it to the nearby bees there in the middle of rural Kenya, just to keep the bees in the loop. They seemed to take the news better than we did.

William Blake once said that the busy bee has no time for sorrow. Standing there under the Southern Cross and a multitude of other stars and constellations, brighter than any stars could be back in Gotham City, I could not remain too sad for long. At some point, my thoughts drifted back to New York and the upcoming beekeeping season. Though it had drawn to an end back there, it would begin again soon enough. Next year's events would unfold in more or less the same predictable rhythm, with some surprises, and most of the same characters would return to take part, with some newcomers thrown in to keep things lively.

I had plans to install an apiary and bee-educational center at the request of the Queens County Farm Museum, a continuously running farm since 1697, the longest continuously farmed land in New York State, right there in Queens, New York City. Also, some wonderful Franciscan friars, part of the Society of Saint Francis, living at the Church of Saint Mary the Virgin, and I were planning an apiary on the roof of their monastery right in the heart of Times Square. In both cases, all of the beehives were being donated by the Durst organization. I was also planning on expanding, giving more classes, workshops, and tours of our urban apiaries, as requests for the same had been so heavy. And I wanted to dedicate more time to a fairly new bee-related passion of mine, which was creating obscure honey bee sculptures made from materials and objects found on New York City Streets, like pieces of broken fire hydrants, wheels from discarded elevator mechanisms, cutlery from shuttered restaurants, broken camera lenses, and so on, cobbled or welded together to at least hopefully resemble a honey bee. There was always more to do and like the busy bee herself, this beekeeper had more than enough to do to stay amused and busy until at least the ripe old age of 106.

Beekeeping in New York City is never boring, between the bees, locations, and wonderful people. Of course, there would be more swarms, and more beekeepers crowding the hipsters and Hasidim in Williamsburg. More aspiring actors, models, and bartenders taking up beekeeping, and more diplomats and retired police officers, too. More laughter with valued and beloved customers at the farmers' markets; more frustration from both two-legged and six-legged creatures; more well-meaning and good-hearted reporters asking the same questions about urban beekeeping while thinking they've discovered something new. More of the organized chaos of the vigorous honey bee colony that is New York City itself.

And I would, should I be so fortunate over the coming years or decades, continue to devote my life to the romance and allure of honey bees, their glorious honey and the vibrant, mysterious society of the hive—while raising my own brood as best I can, and worshipping and tending to, in my way, my own queen.

ACKNOWLEDGMENTS

I am humbled by and appreciate your interest in my story. I had a great deal of direct and indirect help from many people in bringing this book to fruition. The errors are all mine, but any credit also belongs to a wide swath of others. I offer one big blanket apology to those whom I forget to recognize here.

In no particular order: Thanks to my agent, Steve Troha from Folio, who wrote me, then a stranger, asking, "Have you considered a book?" Nine years later, here we are. Back then I was busy building my petty honey fiefdom, shirking my professorial duties, traipsing around the world spreading bee love and getting stung. I could not have imagined dedicating the considerable time it took to get all this down on paper. But Steve doggedly helped make sure that it eventually happened; in fits and spurts we managed, and now the aftermath is in your hands.

My thanks to Pamela Cannon, my editor at Penguin Random House. During the week that Steve took me around to visit the ten publishing houses interested in the proposal, Pamela was our first stop. Though I met a lot of talented people, none held a candle to Pamela. Some proof of that is between these covers. In fact, the entire team at Penguin Random House has been wonderful. To think that my words have been published by the same imprint that brought us J.R.R. Tolkien and Shel Silverstein makes me swell with pride. So thanks to the great team there that includes Amelia Zalcman, Lexi Batsides, Susan Turner, Nancy Delia, Emily Isayeff, and Kathleen Quinlan, among many others. Let me point out that this collaboration has been a lot like the community in the beehive, with me as an outnumbered drone surrounded by strong capable workers.

I must also thank Nan Gatewood Satter for her editorial guidance. She was patient, constructive, good humored, and deft in dealing with my lack of computer literacy. Pamela and Nan have similar editing sensibilities that surely saved me from myself and made this book better than it would have been otherwise. For those who know me personally, if in having read this book my edges seemed less rough; if my words seemed strangely, somehow, softer—do not rejoice or despair. I am, alas, perhaps, as inappropriate as I ever was. I just have talented editors now who tried to disguise that.

Then there are my parents, who have always loved me enough to allow me to make my own mistakes, as frequent as they were, and as painful as they must have been for them to endure. I have been trying to make it all up to them for a long time, and one day I may actually even the score. Also a special thanks to my brother, Michael, who, aside from being particularly helpful in the apiaries during the spring buildup and during harvest times, risks his life daily to protect others, and still has time to be a good father, brother, son, husband, and beekeeper (not necessarily in that order), and has always been there for me. And thanks to his children, Patrick and Megan, who help in a multitude of ways with bee-related matters, and who share their dad with me. I am proud of you all and love you all.

Thanks to my dear friends at Mushin-juku in Kyoto. There are too many to acknowledge all personally so let me thank the matriarch, Ikeda-Sensei, on behalf of everyone for all of the patience, kindness, and affection you have shown me and my family over the last three decades. And a big thank-you to the original Nobuaki for treating me like a son—with a healthy mix of discipline and affection. We all miss you.

Thanks to the Back Yard Beekeepers Association in Weston, Connecticut, where I've had the chance to meet and learn from some of the leaders and great thinkers in the beekeeping industry. I started attending BYBA with my father in the 1980s as a teenager and still go to meetings whenever I can manage. Being asked to speak there in 2006 about my beekeeping experiences in the Middle East was a proud moment for me. The BYBA was the inspiration for the New York City Beekeepers Association, and though we modeled it after them, we will never reach their level of greatness.

My appreciation to the authors who have written about me and my beekeeping undertakings in their own books— including Robin Shulman (*Eat the City*), Alison Gillespie (*Hives in the City*), Leslie Day (*Honeybee Hotel*)—and to Howland Blackiston, who entrusted me to write the urban beekeeping sections for his bestselling book, *Beekeeping for Dummies* (3rd edition forward). Also thanks to the countless journalists who took an interest in me and my work and wrote or filmed about it for audiences all over the world, especially the incomparable Craig Duff.

Thanks to GrowNYC, and specifically the Greenmarket team. As a result of their dedication and hard work, Andrew's Honey has an opportunity to interact with the world at the Union Square Greenmarket. We are grateful to be a part of the impressive collective of farmers and hope to be there for many more years to come. Specifically, many thanks to Michael Hurwitz, Liz Carollo, Laurel Halter, TK Zellers, Jessica Douglas, Margaret Hoffman, Tutu Badaru, Rishma Lucknauth, Aquilino Cabral, June Russell, Lobsang Samten, Cathy Chambers, and Jessica Balnaves for all of the support and kindness, patience and understanding, shown to us over the years. We strive for all sweetness and no stings. I think we have made it!

Specific to our farmers' market stand at Union Square, boundless thanks to my worker bees, present and past: Allison Chan, Hannah Sng Baek, Jen Fraenkel, Sarah Seiler, Heather Rubi, Rose Anderson, Angela Riddlespurger, Susana Yepes, Yangjong Lama, Ingrid Pasten, Mayya Medved Hyatt, and Alyssa Yee, and to drones Benjamin Gardner, Zeke Weber, and Noah Stern, for keeping our customers and friends informed and sweet with our products. Especially Allison, who does most of the work, hand-labeling and painting every single bottle of the New York City rooftop honey, crouched in the back of the converted U-Haul turned honeymobile.

I owe a debt of gratitude to those who have allowed me, past and present, to maintain colonies of honey bees on their properties in New York City: At the Durst Organization, Vanessa Jaworski, Estelle Silberman, and mostly Helena Durst—for their interest, care, and dedication to the precious honey bee; at Brooks Brothers, to Debra and Claudio Del Vecchio, and Emilie Antonetti for their altruistic approach toward beekeeping and all else; to Dr. Robert Koenig, who not only hosts three colonies of honey bees atop the New York Institute of Technology but also allows the NYCBA to hold classes and events in their wonderful space; to Ambassador Peter Thomson and his incomparable wife, Marijcke, without whom the United Nations in New York would not have their stunning North Lawn apiary; to Ann Temkin, Lynda Zycherman, and Glenn Lowry at the unrivaled Museum of Modern Art; to my friend Gus Reckel, who evolved from a banker to become a baker, and now owns L'imprimerie, in Bushwick, Brooklyn, producing extraordinary baked goods sweetened in part by the honey from the three hives I maintain on his roof (the beehives can be seen from the M train, so peek out the window next time you pass the Myrtle–Wyckoff station); to Chef Peter Betz, with whom I have jointly worked the beehives atop the iconic Waldorf Astoria and now at the InterContinental New York Barclay in Midtown; to the folks at Buddhist Insights, who allow me to maintain beehives on the grounds of their monastery for the most enlightened honey I have ever tasted (not that I am too attached to it); to the Clinton Community Garden and Andy Padian, Foram Sheth, and Annie Chadwick, for hosting beehives for the past four decades; to Maggie Christ and Jason Walters at Ballet Tech; to Owen Harrang at Bryant Park; to Jennifer Walden Weprin at the Queens County Farm Museum; to Sibylle Brenner and Schuyler Semlear at Grace Church School; to Ronnie Stewart at York Prep; to Kellie Cahill at the New York Hilton Midtown; Zeke Freeman and the folks at Bee-Raw Honey; and my good friends Kelly Jacques and Gadi Peleg and the whole crew at Breads Bakery; and to all the many community gardens, private rooftops, restaurants, hotels, schools, and other locations too numerous to mention—thanks for your support, for helping the bees to reside in a safe haven, and for your role in enabling these stories to exist.

Continued thanks to the talented beekeepers and educators at Cornell's Dyce Lab for Honey Bee Research, where I am currently trying to finish the master's in beekeeping program. Thanks to Dr. Tom Seeley and Emma Kate Mullen and Shelley Stuart for their help and inspiration.

My great affection and appreciation to Barbara Stein, who, while acting as my academic adviser when I was studying in Kyoto, was the first educator who ever told me that I had a knack for writing. I am glad we are still friends, even though you knew me during my most obstreperous years. Thanks also to John Frederick Ashburne, my subsequent academic adviser in Kyoto and a writer himself. Though he thought I was "a bit of a wanker" then, and referred to me as "Andrew the dog-faced boy," I am glad we are pals now.

Many thanks to those who have encouraged me, provided free legal advice, brought me a bottle of bourbon on a cold market day, helped me find information relevant to this work—or otherwise contributed to this book or the stories herein: friends and family like cousins Peter and Janet Jacobs, Lulu Yee, Hlynur Guðjónsson, Bill and Hope West, Kristi Mackin, Betsy Golden Kellem, Jeanne Laramée, Katherine and Scott Morris, Tom Wilk, Michele Seelinger, Matt Rich, Anna Veccia, Lisa and Rick Agee, Blaine Kruger, Martin Cook, Jo Greenspan, Nils Simon, Heather Fitzgerald, Katherine Gruneisen, Maura Keating, Monica Flores, Simone Drost, Elliot Zelevansky, Gustavo Pita Céspedes, Koji Inoue, Genki Eguchi, and, of course, the Fabulous Pintos, Dennis, Joy, Sasha, and Tristan.

Thank you to big-shot Instagram influencer Eva Chen

(@EvaChen212) who posts (alas not exclusively) about my honey and beekeeping—especially the matcha-infused honey. Until recently I had no idea what an influencer was, but I am glad that you like what I do and what my bees make, and that you tell others about it. It is fun learning about your world. Thank you for not using your considerable powers for evil. Keep spreading sweet love. I hope Ren and Tao finish their lunches today and that Tom perfects his granola recipe.

Sincere thanks to the beekeepers and beekeepers-to-be with whom Bees Without Borders has worked in all corners of the world. We will keep trying to elevate others through bee-keeping endeavors.

Most of all, of course, I want to thank the loves of my life: Yuliana, who for reasons known only to her, has not yet absconded from me, and to my much treasured son, Nobuaki. There is no word or sentiment in any language strong enough to convey the depth of my love for them. Thanks for giving me the space and encouragement to write this book, even though, thinking back, I wrote most of it in early mornings while you were both still sleeping, so, frankly, I am not sure if either one of you should get too much credit. But thanks for being there.

And though they do not read, and worse still, do not buy any books, I wish to go on record as thanking the honey bees themselves. They are magical creatures, and on the backs of their wings I am able to cobble out a living doing something I love. So thanks, bees, for tirelessly and selflessly working your-selves to death so we can all sweeten our tea and saturate our toast with your vomit.*

---

* Honey isn't really bee vomit, but this is poetic license.

ANDREW COTÉ is New York City's most well-known beekeeper. A fourth-generation apiarist, he is founder of the New York City Beekeepers Association, executive director of the nonprofit Bees Without Borders, polyglot, Fulbright scholar, black belt in aikido, and former college professor. Andrew and his bees have been featured on *The Martha Stewart Show,* CNN, CBS, *Cake Boss, Dr. Oz, Nightline, Good Morning America,* and *Today,* and in *The Wall Street Journal, The Atlantic, The New York Times, HuffPost,* and many other news outlets around the world. Andrew also runs Andrew's Honey out of the NYC Union Square Greenmarket.

AndrewsHoney.com
Instagram: @AndrewsHoney

ABOUT THE TYPE

This book was set in Baskerville, a typeface designed by John Baskerville (1706–75), an amateur printer and typefounder, and cut for him by John Handy in 1750. The type became popular again when the Lanston Monotype Corporation of London revived the classic roman face in 1923. The Mergenthaler Linotype Company in England and the United States cut a version of Baskerville in 1931, making it one of the most widely used typefaces today.